Historical Geography and Geosciences

This book series serves as a broad platform for contributions in the field of Historical Geography and related Geoscience areas. The series welcomes proposals on the history and dynamics of place and space and their influence on past, present and future geographies including historical GIS, cartography and mapping, climatology, climate history, meteorology and atmospheric sciences, environmental geography, hydrology, geology, oceanography, water management, instrumentation, geographical traditions, historical geography of urban areas, settlements and landscapes, historical regional studies, history of geography and historic geographers and geoscientists among other topically related areas and other interdisciplinary approaches. Contributions on past (extreme) weather events or natural disasters including regional and global reanalysis studies also fit into the series.

Publishing a broad portfolio of peer-reviewed scientific books Historical Geography and Geosciences contains research monographs, edited volumes, advanced and undergraduate level textbooks, as well as conference proceedings. This series appeals to scientists, practitioners and students in the fields of geography and history as well as related disciplines, with exceptional titles that are attractive to a popular science audience.

If you are interested in contributing to this book series, please contact the Publisher.

Christian Sellar · Gianfranco Battisti

Geopolitical Perspectives from the Italian Border

Introducing Gianfranco Battisti, Triestino Geographer

Springer

Christian Sellar
Department of Public Policy Leadership
University of Mississippi
University, MS, USA

Gianfranco Battisti
University of Trieste
Trieste, Italy

ISSN 2520-1379 ISSN 2520-1387 (electronic)
Historical Geography and Geosciences
ISBN 978-3-031-26043-8 ISBN 978-3-031-26044-5 (eBook)
https://doi.org/10.1007/978-3-031-26044-5

This Springer imprint is published by the registered company Springer Nature Switzerland AG
The registered company address is: Gewerbestrasse 11, 6330 Cham, Switzerland

To our wives, Shuang and Tiziana

Acknowledgements

We owe a debt of gratitude to the people that made this book possible. We are especially thankful to Prof. John Pickles, University of North Carolina at Chapel Hill, for his feedback on the initial book concept; Prof. Michele Stoppa, University of Trieste, for drawing Figs. 7.1 and 11.1; Prof. Giovanni Mauro, University Vanvitelli, for drawing Figs. 10.1 and 10.2; Ms. Doris Bleier, Editor of Earth Sciences, Geography, and Environment at Springer Nature, for her incredible patience and guidance; an anonymous reviewer, for the constructive comments and positive feedback; and our families, for their enduring support.

Contents

Part I Introduction

1 Writing Geopolitics in Trieste: Gianfranco Battisti's Intellectual Production in the Context of the 'Geography of Geographies' ... 3
- 1.1 Introduction: Bringing a Triestino Geographer to an International Audience ... 3
- 1.2 Intellectual Contribution ... 5
- 1.3 Methodology ... 6
- 1.4 The Chapters ... 7
- References ... 11

2 Geography as Industry: The Institutions, Untraded Interdependencies, and Worlds of Production Shaping Academic Work ... 13
- 2.1 Introduction: The Professional Framework of Geographical Thought ... 13
- 2.2 Institutions as Mutual Expectations ... 14
- 2.3 Rethinking 'Untraded Interdependencies' and 'Worlds of Production' Through an Institutional Lense ... 16
- 2.4 Toward an Academic World of Production ... 18
- 2.5 Conclusion: The Benefits of Viewing Geography as an Industry ... 20
- References ... 21

Part II Triestino Geographical Thought

3 Becoming a Geographer in Trieste. Autobiographical Essay, Reflecting on the Nature of Geography ... 25
- 3.1 Joining the Institute of Geography ... 25
- 3.2 Understanding Triestino Geography ... 27
- 3.3 Developing My Vision of the Discipline ... 29
- References ... 30

4 **Living in the Borderlands: Political Geography, Geopolitics, and Advocacy in the *Triestino* School of Geography During the Long Twentieth Century** 33
4.1 Introduction: The Civic Commitment of *Triestino* Intellectuals 33
4.2 The Interwar Period: The Formative years of Triestino Geography 35
4.3 After WW2: Geographers' Policy Activism in a Bipolar World 37
4.4 Falling Walls: Triestino Geography at the End of the Bipolar World 40
References 42

Part III Historical Geography as Method: Producing Geopolitics from the Julian Region

5 **Urbanization Processes in a Transnational Area. An Application of the Rank-Size Rule to the Austrian Littoral** 47
5.1 Introduction—Some Issues with Territories 47
5.2 The Area Under Investigation 48
5.3 Methodology 49
5.4 Regional Dynamics 51
5.5 Data Analysis 52
5.6 Conclusions: The Many Reasons of Places 56
References 57

6 ***Gorizia Nova*, Aka "New Gorizia:" A Euro-City on the Border Between Italy and Slovenia. A Recommendation for Local-Level Territorial Changes After Slovenia Joined the European Union in 2004** 59
6.1 Introduction: Integrating the Two Gorizias 59
6.2 The Territorial Challenges of Gorizia Nova 61
6.3 The Identification of the Areas 62
6.4 Defining *Gorizia Nova*'s Urban Core 63
6.5 Conclusion 66
References 67

7 **Inland Areas and Border Regions: A Geopolitical Interpretation. Comparing the Marginalization of Trieste and Umbria as Examples of the Dynamics of Borderlands Versus Remote Inland Areas** 69
7.1 Introduction: The Marginality of Internal Areas 69
7.2 Geographical Marginality and Economic Marginality 70
7.3 The Inevitable Reversibility of Geographical Conditions 72
7.4 Borders as "Actors" of Regionalization 74
References 75

Part IV From Trieste to the World: Deploying Triestino Geographical Thought to Grand Geopolitics in Europe and Beyond

8 The Reshaping of German-Yugoslav Space from a Middle European Point of View. Paper Presented at the 2nd IBRU Conference, Held in Durham, UK, on July 18–21, 1991 79
8.1 Introduction: Reading Thirty Years Old Notes 79
8.2 Paper Presented to the 2nd I.B.R.U. Conference (Durham, 18–21 July 199I). Foreword 80
8.3 The Yugoslav Question. 81
8.4 Toward a New Danubian Confederation? 82
8.5 New Problems for Old Minorities 82
8.6 Is the German Question Really Ended?. 84
8.7 Will the Soviet Union Disintegrate? 85
8.8 The Role of the European Community 87
References. 87

9 Europe. The Many Reasons of an Epoch-Spanning Crisis. A Long-Term Geohistorical and Geoeconomics Analysis of the Obstacles to European Integration 89
9.1 Introduction: A Long-Term Analysis of European Geopolitics 89
9.2 Between History and Geography. 90
9.3 The Institutional Question 91
9.4 From the European Communities to the Union 93
9.5 The Weight of History 94
9.6 Towards a Pluralistic Europe. 96
9.7 A Continent in Disarray 97
9.8 A Gaze in Retrospective 98
9.9 The Perspective of the Economy. 99
9.10 Conclusions. 100
References. 101

10 A Century of Struggles. A Comparison of Multiple Geopolitical Agendas in Europe, the USA, and Beyond 103
10.1 Introduction: The Repetitiveness of Geopolitical Nodes 103
10.2 The Problem of Primacy 104
10.3 Divide and Rule 105
10.4 American Crises 106
10.5 The Strategic Framework 108
10.6 A Preemptive War 110
10.7 The Resurgence of Russia. 112
10.8 Europe in Disarray 113
10.9 The Many Problems of an Emerging Superpower. 114
10.10 Escalating the Crisis 116
10.11 Conclusions. 117
References. 118

11 Cycles of Geopolitical (Dis)order, as Determined by Interactions Between Spatial Systems. A Theoretical Model of the Systemic Drivers of Geopolitics 121
11.1 Introduction: A Systemic Approach to the Geopolitics of Conflicts 121
11.2 The Role of Conflicts 123
11.3 Globalization in Waves 124
11.4 Toward the End of a Wave 127
11.5 Conclusions 128
References 129

Appendix: Gianfranco Battisti's Thematic Bibliography 131

Index 137

List of Figures

Fig. 3.1 University of Trieste main building: historical site of the Institute of Geography (1961). *Source* University of Trieste archive ... 26

Fig. 4.1 Two centuries of borders in the upper Adriatic area. *Source* Barbi (1975) ... 34

Fig. 4.2 Italian-Yugoslav cross-border areas with facilitated commercial traffic: clearing agreements for Gorizia/Nova Gorica and Trieste/Capodistria (Koper). *Source* Battisti (1984) ... 39

Fig. 4.3 Gorizia: a "cold war" border. *Source* ModelplaninterregIIIa ... 41

Fig. 5.1 The Austrian Littoral province. *Source* ModelplaninterregIIIa ... 49

Fig. 6.1 Gorizia area: local governments on the Italian-Slovenian border. *Source* ModelplaninterregIIIa ... 60

Fig. 6.2 GoriziaNova functional areas. *Source* ModelplaninterregIIIa ... 65

Fig. 6.3 Gorizia Nova: a tentative framing. *Source* ModelplaninterregIIIa ... 66

Fig. 7.1 Border regions. **a** The logic of politics. **b** The geographical reality. *Source* original elaboration ... 71

Fig. 7.2 Italy, showing the regions and provinces discussed in the chapter. *Source* original elaboration ... 72

Fig. 8.1 Geopolitics in progress: the Yugoslav drama. *Source* original elaboration ... 83

Fig. 8.2 Patterns of uncertainty on the Baltic shores. *Source* original elaboration ... 86

Fig. 10.1 Geopolitical associations. *Source* original elaboration ... 106

Fig. 10.2 Geopolitical protagonists. *Source* original elaboration ... 117

Fig. 11.1 The space model of contemporary globalization. *Source* original elaboration ... 125

Part I

Introduction

Writing Geopolitics in Trieste: Gianfranco Battisti's Intellectual Production in the Context of the 'Geography of Geographies'

1

Abstract

Why should international audiences care to read the work of Gianfranco Battisti, an Italian geographer based in the port city of Trieste? In the context of the intellectual debates over the inclusion of non-English-speaking geographers in leading international publications, Battisti's work provides several insights. These include: (a) fresh perspectives on geopolitics and globalization from the point of view of a city and region that was both a Cold War hotspot and a laboratory of post-Cold War institutional experimentation; (b) a window on an intellectual tradition that drew on the same nineteenth century classics as international geography but developed in a relatively autonomous and original way; (c) a challenge to the established dichotomy between realist and critical approaches; and (d) a reflection on the institutional nature of knowledge production, which occurs within constrains of career selection, development, and incentives not too dissimilar to those experienced in other industries. To showcase his work, the authors arranged this anthology around three overarching themes: the specificities of Triestino's geographical thought, the geopolitics of Trieste and surrounding areas, and Europe-wide and global trends.

Keywords

Translation · Inclusion · Geographical thought · Geographical traditions

1.1 Introduction: Bringing a Triestino Geographer to an International Audience

Early in the millennium, human geographers began to pay attention to translation, i.e. the transmission of different forms of knowledge across language and spatial boundaries. They questioned the extent to which the knowledge they produce is truly international (Gutiérrez and Lopez-Nieva 2001; Kitchin and Fuller 2003; Rodríguez-Pose 2004, 2006). The first, and obvious, critique has been that the so-called international geographical journals are indeed British and American, and foreign scholars rarely get published unless they hold Ph.D.s from universities in the USA or UK (Aalbers 2004; Aalbers and Rossi 2007). Theoretical pieces that build upon non-Anglo-American scholarly traditions are rarely recognized as meeting the standards of international geography (Minca 2000). Responding to this critique, several journal special issues have focused on the need to

C. Sellar and G. Battisti, *Geopolitical Perspectives from the Italian Border*,
Historical Geography and Geosciences, https://doi.org/10.1007/978-3-031-26044-5_1

open up international human geography to the contributions of non-English-speaking scholars and traditions (*Environment and Planning D* 2003; *European Urban and Regional Studies* 2004; *Geoforum* 2004).

In the following decade, scholars have continued analyzing the 'geographies of Geography,' i.e. the spatialities of geographical knowledge production (Jazeel 2016). On the one hand, they somberly noticed that the leading international journals are still overwhelmingly publishing authors affiliated with universities in the global North—93% in *Antipode*, despite the radical leaning of the journal, and 95% in *Progress in Human Geography* in 2013 (ibid.). On the other hand, they are problematizing the hegemonies and counterhegemonies of knowledge production on the backdrop of persisting 'national' geographical tradition at a time of heightened scholarly mobility. First, the conditions of mobile (migrant) scholars studying communities outside the 'EuroAmerican heartland' are increasingly scrutinized (Gawlewicz 2016; Jazeel 2016). Second, the institutional frictions involved in academic mobility are becoming a topic of interest (Van Riemsdijk and Wang 2016; Jöns 2018). In this respect, Heike Jons highlights the persistent relevance of 'national' geographical traditions, concluding that "despite the age of the Internet, human geographical debates remain powerfully structured by different academic languages, and by nationally orientated research traditions, publication cultures, promotion criteria, and networking opportunities" (Jöns 2018, p. 32).

In discussing the long-term shift in hegemony between German and English-speaking geographies, Jöns noticed the various level of 'stickiness' and permeability of each tradition. In so doing, she presented anecdotal cases of circular cross-pollination, in which, for the most part, young scholars traveled to the hegemon to 'import' new ideas in their domestic, non-hegemonic, contexts (2018, p. 30). The third strand of research on the geographies of geography works against this backdrop of unidirectional knowledge transfer, with vigorous attempts to present distinct traditions to the wider world—such as, for example, Simonsen and Öhman's work on Nordic geographies (2018). In so doing, some geographers are trying to reverse the flow of knowledge by bringing ideas from the 'South' to international geography, such as, for example, the symposium in *Antipode* dedicated to the work of the Brazilian geographer Milton Santos (Melgaço 2017).

This book contributes to the latter line of research by synthesizing for an international audience the work of Gianfranco Battisti, an Italian geographer writing from the backdrop of Trieste, a port as well as the largest city in a border region that played a key role first, in Cold War geopolitics, and, later, as a laboratory of institutional restructuring during the eastward enlargement of the European Union. At different points in time, Trieste was the capital city of French, Austrian, and Italian provinces, governing an ethnolinguistic and politically fractious area surrounding the *Julian Alps*, the mountain range that today marks the Italian/Slovene border. Thus, writing from Trieste means both empirically focusing on some of the geopolitical flashpoints of the last two centuries, and theoretically reflecting on the meaning and nature of geopolitical 'facts.' All chapters are original publications, written for an international audience to capture nearly fifty years of intellectual work. As such, they build on a variety of earlier papers, part of a prolific and multifaceted intellectual production started in 1971 and continuing to the present. The appendix lists the author's publications closest to the themes discussed in each chapter.

This book is also the fruit of a friendship and mentorship developed in more than 25 years. To emphasize the reflexive nature of this work, the authors decided to use first person pronouns when meaningful. Each chapter includes the author's name to clarify whose 'voice' is speaking. In this chapter and the following (Chap. 2), 'I' refers to Christian Sellar; in the remainder of the book (Chaps. 3–11), 'I' is the voice of Gianfranco Battisti. Throughout the book, 'we' reflects our conversations over the years as well as our collective voice in the coauthored chapters (Chaps. 6, 9, and 11). In so doing, we meant, on the one hand, convey to the audience the respect of a student to his former teacher. Battisti was my undergraduate advisor at the University of Trieste; he then recruited me in the Ph.D. program in *geohistory and geoeconomy of border regions* in the same university, to then let me go to pursue another Ph.D. in geography at the University of North Carolina at Chapel Hill, under the supervision of John Pickles. Conversations with Gianfranco and John in the Spring of 2001 and the preparation of a Ph. D. application in the USA in the following fall—during the craze following the September 11th's terror attacks—shaped my life to an unlikely path: I went from a small town in Italy's borderlands to making a home in the American heartland. Professionally, my research focuses on the boundaries between political and economic geographies, on the "geopolitics of economic geographies" (Sellar 2019). Thus, simply put, I am comfortable introducing to an international audience Battisti's work closest to my own. On the other hand, we also meant to synthesize a decades-long research program, bearing two goals in our minds: first, to provide a coherent presentation of his approach to geopolitics, and second, to make it more accessible to its intended audience. Thus, the book is organized to maximize contextual information, walking the reader through Battisti's biography and environment to explain the rationale behind the production of unconventional theories, and then adapting these to contemporary discussions over sovereignty, European integration, and globalization.

1.2 Intellectual Contribution

In translating Gianfranco's work, the first question we asked ourselves was about the audience. What exactly is international geography? Is our aim to speak to all geographers outside Italy, or to a subset? Who should we cite to contextualize his work for a wider audience? In the end we settled for the Anglo-American core of the discipline. Bearing in mind Jöns observation about the stickiness of national traditions in mind (2018, p. 32), we focused on the dialogical and dialectic nature of intellectual work. Simply put, we choose to traditions and schools of thoughts as groups of scholars that engage and know each other through conferences and through reading each other work. Once we foregrounded the community dimension of intellectual work, bounded by language and training if not strictly by nationality, it became clear that targeting Anglo-American geographers has two major benefits. First, it made easier to identify theoretical debates. For example, the last section of this book asks questions about the interplay between sovereignty, geopolitics, and geoeconomics that fits well with works by John Agnew (1994), Matthew Sparke (2018), and Moisio (2019) and others. John and Matt are both based in the USA; Sami is a Finn whose publications often land in highly ranked Anglo-American journals. Second, it made it easier to reach a wider audience, because these English-speaking scholars are likely read by the larger international community. In a nutshell, coherence in intellectual debates and wider reach led us to focus on engaging mostly Anglo-American geographers.

Once we determined our audience, we had to answer the key question asked to all translations. As Melgaco aptly put "but why should I know him? There may be thousands of interesting geographers unknown to the English-speaking community, so why should one care about this specific author?" (2017, 946). The first answer is thematic. Simply put, Trieste is an interesting place for the student of geopolitics. This city was a site of struggles in the emergence of nineteenth-century nationalism; later, Italy's territorial gains after WW1 and losses after WW2 placed the city right on the old Iron Curtain and made it the first flashpoint of the Cold War (Jennings 2017). In the next historical upheaval, post-Cold War debates on European integration led to the scholarly reinterpretation of Trieste in the framework of cosmopolitanism (Waley 2009). Overall, Trieste and the neighboring areas provided empirical materials in framing scholarly debates over European identities (Bialasiewicz 2009) as well as inquiries in critical geopolitics over deterritorialization and reterritorialization (Newman 1998).

The second answer to Melgaco's question is conceptual, aimed at adding nuance to the discussion over the geographies of geographical knowledge production. As Melgago himself put it, the very question of inclusion is in itself a reflection of hegemony, posed to outsiders wishing to 'join the club.' On the other hand, the condition of hegemon implies producing terms of reference that others rarely question. For example, remaining with the Italian case, articles recently published in the leading journal *Rivista Geografica Italiana* habitually cite domestic scholars, former hegemons (mostly German and French scholarship), and current Anglo-American scholars (Mayoral and Curiazi 2019; Minca 2019). Thus, while the non-hegemon may be able to draw from multiple sources, the hegemon may be tempted to frame a simple dichotomy between itself and 'the other.'

As in the seminal work by Minca (2000) and the more recent efforts by Jöns (2018), this collection aims at problematizing 'the other' by highlighting the distinctiveness among various traditions. First, much of the normative conversation about inclusion in geography occurs along with the binaries of privileged/marginal, which in turn is rooted in the progressive agenda of researchers driven by postcoloniality (Gikandi 2001; Jazeel 2016). Similarly, to Simonsen and Öhman (2018) discussion of Nordic geographies, our focus on the Italian case reflects an intermediate case of partial exclusion, of being definitely outside the core of international geography, but close enough to entertain a subaltern, but still active conversation (as in Minca's work). Second, this anthology suggests looking at a finer grain than the implicit methodological nationalism that defines geographical traditions as articulated around language communities and common institutional structures (Jöns 2018). Instead, this anthology suggests looking at the urban as well as transnational specificities of the *Triestino* school of geography, much influenced by conversations about borders, and by the sometimes fractious, sometimes cooperative relations with its Slovenian neighbors (Chap. 4 in this collection).

The third answer is the theoretical contribution to geopolitics. Evolving from nineteenth- and early twentieth-century French and German classics—Vidal de la Blache, Ratzel, and Christaller—that were once influential on Anglo-American geography, *Triestino* geographers developed following their path-dependent trajectory. These developments displayed a certain level of originality, which includes an unexpected blend of realism and critical approaches to geopolitics; a lack of dismissal of the quantitative revolution, and the evolution and continuation—rather than dismissal—of twentieth-century classics such as Christaller and Loesch. In a nutshell, while considerable portions of Anglo-American human geography proceeded in a series of upheavals, accompanied by critiques of earlier paradigms (Livingstone and Godlewska 1992; Massey et al. 1999; Gould and Pitts 2002), Battisti's work emphasizes continuity. To the Anglo-American reader, the theoretical implant may look outdated, but the resulting analysis provides useful insights on debates on the relationship between geopolitics/geoeconomics (Moisio 2018), sovereignty regimes (Agnew 2009), and a provocative viewpoint on the end of the Cold War, Europeanization, and American imperialism.

The fourth answer is a contribution to the history of geographical thought. In discussing the development of the discipline, we rarely see ourselves as an *industry*—i.e. and organized activity aimed at producing a specific product (geographical knowledge), with diverse modes of production, and also governed by national institutions, and dependent on regional untraded interdependencies that shape path dependencies (Storper 1997). A closer look at Battisti and his tradition is a mirror that may reveal something new about the implicit rules and path dependencies that frame the international geographical tradition, highlighting our strengths, but also, in some cases, showing the blind spots that any system by necessity generates.

1.3 Methodology

Gianfranco provided me either with full chapters or notes in Italian, which I reworked in translation to come closer to the expectations of international academic readers. We agreed on writing original pieces, which also reflected, either in themes or methods, his earlier work in Italian. To highlight the links between this and his earlier work, this book includes an appendix with his publications in geopolitics and political/economic geography, reporting in bold characters the titles more closely related to this book.

Once I received the chapters, first of all I worked through their logical flow. Italian academic writing tends to build arguments on a syllogistic structure, beginning with premises and reaching a conclusion toward the end of the piece. When appropriate and respectful of the original content, I reversed the logic introducing concise arguments at the beginning of the chapters, to bring them to a structure more familiar to the intended audience. Then, I used an online translator to provide a rough draft to build on. After comparing two different translators, I opted for DeepL Translator. Then I combed the text looking for statements, arguments, and information that may not be familiar to international readers. I made those explicit, omitting sections that may have not been relevant for audiences outside Trieste and Italy. When necessary, I updated obsolete information. In consultation with Battisti, I introduced references to theoretical debates in international geography sparingly, only when they made the arguments clearer to our intended audience. Finally, I checked the grammar and structure of sentences and paragraphs. I concluded with further electronic verification of grammar and syntax using the Grammarly app. The chapters single authored by Battisti were good standalone papers in their original Italian version; the coauthored ones include my substantive contributions. Both authors revised the final text.

1.4 The Chapters

The chapters are arranged in thematic sections. This first introductory section (This chapter and Chap. 2) aims at placing Battisti's work in direct dialogue with key themes in international human geography. The next chapter insists on the notion of geography as an industry. It aims at explaining Gianfranco's work as the product of specific institutional constraints in his profession (Italian academia) and workplace (the University of Trieste), arguing that true inclusivity must include an appreciation of the diversity of incentives academics experience in their workplaces. The chapter includes a discussion of Michael Storper's work on path dependencies in industry, which included the observation that certain Italian economies were strong producers of 'follow up innovation'—i.e. improvements over existing technologies—rather than cutting edge new technologies (Storper 1997). The notion of follow-up innovation captures well the nature of Battisti's intellectual contributions. During the twentieth century, Triestino geographers imported French landscape studies, German spatial economy, the Anglo-American quantitative revolution, and, more recently, attention to issues of Europeanization and globalization. Rather than attempting radical theoretical innovation, in nearly four decades of scholarship, Battisti used and refined all these approaches, emphasizing the elements of continuity among them.

Theoretical continuity has led to two outcomes. First, drawing directly on French landscape studies, Battisti treats historical geographies as a method, actively using long-term historical analyses to better understand contemporary geopolitics. Taken together, this book's chapters show the mutual influences between historical experiences—reflecting in shared, albeit contested, identities, providing a backbone for political unity, which in turn determines the success (or failure, as in Chap. 5) of economic and political strategies. Second, attention to the historical processes that shape geopolitics addresses the tensions highlighted by studies of the territorial trap (Agnew 1994), as well as the competing visions of states as static frames or dynamic networks (Moisio 2013). In a nutshell, Battisti argues that, by viewing territorial states from a long-term perspective, scholars can portray them as keystone systems dedicated to maintaining stability and consensus that enables other social and economic systems to operate (Chap. 11). In his view, interstate conflicts are flashpoints generated by rapid changes and loss of equilibrium in a variety of social systems that have network-like spatial extensions seldom coinciding with one another and with state boundaries. Thus, Battisti places his scholarship in an intermediate position between realist geopolitics—which views states as units of analysis of the international system—and state theorists that privilege views of states as strategic relations among actors (Jessop 2001).

The second section, comprising Chaps. 3 and 5, sets the stage by introducing the author and his school of thought in their urban context. Chapter 3 is an autobiographical essay, similar in style to the *Fourteen Geographical Voices* edited by Gould and Pitts (Gould and Pitts 2002). The essay shows the continuity with a geographical tradition predating Anglo-American hegemony, vividly including a quote from his mentor, defining geography as a 'difficult discipline' starting with cartography, developing physical geography in conjunction with the history of human presence in the land, and transforming economies and societies. Only after all the above is completed, 'the real scientific work can begin' (p. x). Such an approach echoes the nineteenth-century German geographer Karl Ritter. Ritter methodological influence did not happen by chance: in Trieste as elsewhere, German geography was the hegemon before the UK and USA. Such influence extended to the point that the first question Battisti was asked when recruited to work in the department was 'do you speak German?'.

Successful or not in answering that particular question, Gianfranco joined the University of Trieste in 1971, not to leave until his retirement in 2016. He was trained during the quantitative revolution, which influenced him as much as the earlier landscape studies. From the former, he acquired a methodology based on hard and measurable data (and, I would add, a certain antipathy toward the most abstract forms of postmodernism); from the latter, he derived a passion for historical analyses in which the physical, social, and economic contingencies in a bounded territory provide data for each specific inquiry. Such methodological preferences shape Chap. 4, in which he discusses the contribution of the *Triestino* school of geography tying scholars' intellectual pursuits with their institutional context, in turn, shaped by the city's position in twentieth-century grand geopolitics. Geographers—in their capacity as cartographers, students of the landscapes, of trade flows, and local politics—had to become public intellectuals, so they could help their city weather the momentous upheavals of the twentieth century.

Simplifying local history for readers unfamiliar with European historical geography: Trieste had attracted considerable wealth in the nineteenth century because it was the most important port city of the Austro-Hungarian Empire. Later, in the twentieth century, the city's fortunes were altered twice: first with the annexation to Italy after WW1, and then a second time after WW2, when Italy's territorial losses to Yugoslavia threatened the territorial integrity of the city while turning it into a Cold War frontier (Giraldi 2016). These border changes meant that Trieste lost its role as an international trading hub after WW1, because Italy, unlike Austria, had a long coastline and plenty of port cities. In such conditions, it soon experienced economic decline, only in part compensated by investments from the Italian government. Twenty years later, WW2 engulfed the city. At the end of the war, it was occupied for forty days by Yugoslav forces and threatened with ethnic cleansing. To avoid further escalation, the USA and UK occupied it for a few years, then returned it to Italy. At that point, a new international border between Italy and Yugoslavia was established in Trieste's backyard. The 1950s were challenging, as the new international border cut the city off its market and supply lines for agricultural products, while at the same time causing an inflow of refugees in need of both housing and jobs. Also that border was not just any border, it was the Iron Curtain where East and West collided. In such a context, geographers were pretty much forced to work beyond the ivory tower to aid the survival of the city. The chapter discusses the specific organizations designed to channel their efforts. Their work focused on the political-economic geography of border regions because they were preoccupied with political issues (such as administrative reforms after WW1, reconstruction after WW2, the Marshall Plan, the relationship with Yugoslavia), but only to the extent to which they facilitated or hindered trade flows, access to raw materials, maintained the physical landscape and protected the transportation route with Italy's mainland. As public intellectuals fighting for the survival of the community, key *Triestino* geographers established relations with their Slovene colleagues across the border, anticipating a trend of cooperation that could fully flourish only after Yugoslavia and Italy signed a formal peace treaty in 1976, the 'Treaty of Osimo.'

The third section includes Chaps. 5–7, focused on Battisti's work on the Julian region. His approach is influenced by the policy advocacy of the *Triestino* school, which included a focus on the strategies and consequences of governments drawing the boundaries of subnational administrative units. Thus, Chap. 5 discusses the evolution of administrative units around Trieste. Taken together, the chapters showcase long-term historical thinking, a theoretical approach based on twentieth-century classics—Loesch and Christaller–but also analytical rigor that does not shy away from spatial statistics and GIS. Chapter 5 discusses the resilience of subnational administrative units, by looking at the long-term legacies of the nineteenth-century Austrian province within which Trieste represented the top of the urban hierarchy. The chapter compares census data from the nineteenth and twentieth centuries using the rank-size rule, one of the earliest quantitative methods used in urban geography. Chapter 6 is set against the backdrop of the 2004 enlargement of the European Union, when Slovenia joined the EU and opened its border with Italy. The chapter discusses the reintegration of the urban space of Gorizia, a nearby town split in two during the Cold War. Using GIS techniques, the chapter discusses the appropriate scale of joint urban planning across an international border, given that Italian and Slovenian subnational administrative units are hardly comparable in both size and criteria used to distinguish rural and urban areas.

Driven by such focus on a city and region where the frequent changes in borders threatened the economic foundations of the city, Battisti anticipated the view, currently pushed by Moisio and others (Moisio 2018; Sellar 2019), that geopolitics and geoeconomics are tightly interconnected, or even the same analytical category (Moisio 2019). In a short conceptual essay, Chap. 7 compares border regions and inner regions, arguing that the dynamic between economic development and underdevelopment across Europe is largely driven by centuries-long geopolitical dynamics caused by wars and shifts in borders. At the same time, more than contemporary work in geopolitics/geoeconomics—both critical and realist—he sees wars and conflicts as connections between geopolitics and geoeconomics, building directly against the argument that geoeconomics started as a reaction against traditional geopolitics (Luttwak 1990; Cowen and Smith 2009; Sparke 2018; Vihma 2018).

The fourth, and last, section of the book is dedicated to Battisti's approaches to globalization. Throughout his career, but more frequently in his later years, he paid increasing attention to the mutual feedbacks of conflicts, geoeconomics, and geopolitics, scaling up his analysis from the local to the European and global scales. Europe-level conflict is the theme of Chap. 8. In an exploratory conference paper delivered at Durham University in the UK in 1991, he predicted the upcoming violent breakup of Yugoslavia, outlined the risk of conflict spreading to Central and Eastern Europe, and anticipated the role of the European Union in deescalating tensions. The paper was rejected from publication at the time, as it violated quite a few unspoken norms in Anglo-American geography: we usually refrain from prediction, we either base research on primary data, or the production of theory, or both. The chapter instead built on good use of some descriptive statistics, literature from Yugoslav scholars, and—most important—a keen eye to political news. In retrospective, quite a few of his predictions came to pass, showing that the rules of international academia, although extremely effective, are not immune from blind spots that 'others' may be able to fill.

The following Chap. 9 uses Battisti's preferred method—historical geography as a tool for analyses aimed at unveiling the mutual influences between history, realist geopolitics, and geo-economics of a given space—to perform a long-term analysis of the evolution of European territorial states to reveal some of the vulnerabilities of the European Union as a project of continental integration. More than other chapters, this piece reveals the author's political and philosophical ground close to the classics of Anglo-American conservatism, with a position on the French Revolution, the nature of modern nation-states, and the role of religious faith in public life that is reminiscent of Burke (2008), Babbitt (1924), and Kirk (2005). Albeit a conservative and a realist, the author does not fall into the territorial trap of considering nation-states as a natural, pregiven reality (Agnew 1994, 2010). Instead, the various form of political power are treated as dynamic spatial systems that can change, qualitative as well as quantitatively, over time following institutional and economic circumstances. Thus, the chapter is nevertheless able to break the box of territorial thinking by viewing the territorial state as an adaptable, context-specific mean to an end, rather than as a fixed container of social relations.

In the last two chapters, Gianfranco brings his history-driven, geopolitical thinking to the global scale. Chapter 10 brings the analysis of the previous chapter from the European to the global scale, discussing the changing dynamics of the state system in the century following World War 1. In discussing shifts in hegemony, the piece discusses the interplay of cooperation and competition within the Western world, driven by the relationship between the European Union and the USA. Continuing his reflections on how geopolitics is ultimately framed by the need to create consensus around economic practices, the chapter emphasizes the goal of the European single currency, the Euro, as a barrier against the seigniorage of the US dollar. In so doing, the chapter offers a continental European viewpoint on the discussions on the relations between the US dollar as the currency of global trade and American hegemony (Gowan 1999; Ivanova 2010).

Chapter 11 concludes the book by tying together his earlier reflections on the interactions between historical processes leading to and economics-framed geopolitics and globalization. It starts with a nod to realism, by treating legally recognized territorial states as the units of a global system, and defining geopolitics as the spatial relations among such units. However, the chapter also takes stocks of a variety of underlying systems (cultural, economic, and even institutional) that tend to flow across states' borders. The territorial state is the most important element in this 'system of systems,' because the political sphere is tasked with maintaining the consensus necessary for the other spheres to function. States must manage changes in the other subsystems while maintaining consensus, and thus, any political crisis at the state level is both an indicator and a cause of conflicts in other subsystems—the hardest to manage to be multi-systemic crises. For example, the current (in the spring of 2022) conflict in Ukraine is particularly difficult to resolve partly because it is rooted in the convergence between competing economic structures and divergent cultural spheres.

While international geography tends to see a direct opposition between critical and realist approaches (Vihma 2018; Sparke 2018), Battisti seems to give a nod to both. First, as the realists, he sees geostrategic competition among states as a key element of analysis; however, he resists the temptation to configure his work as a tool for policymakers, as they tend to do (Luttwak 1990). In so doing, he tackles the main complaint brought by critical geographers to realists, i.e. that such 'strategic' (aka policy-oriented) thinking is a form of theory with dubious political implications (Moisio 2019, p. 3). In a nutshell, critical geopolitics aims at speaking truth to power, and so does Battisti. Second, he is as attentive to conflict as the realists, but he refuses to reduce geopolitics to a narrow analysis of military power (ibid. p 4). Instead, military conflict is interpreted as one among many outcomes of disequilibrium among spatial relations in a variety of subsystems that—on the one hand—constitute the state, and—on the other, are not limited by borders. Like critical scholars

(Sparke, Toal, Moisio, etc.), he does not take the state for granted. Instead of seeing it as a coherent agent, his understanding of multiple systems in relationship with each other is compatible with state theorists, in particular with those pushing for a strategic-relational view of the state (Jessop 2001; Brenner 2004). In sum, by developing further an academic tradition indigenous to his city and country, Battisti was able to build approaches that complement some of the boundaries and binaries that shape international geopolitics—and geography more in general. I found the journey through the intellectual work of my mentor truly fascinating, and hope to convey to readers the pleasure of discovery I felt while writing this book.

References

Aalbers MB (2004) Creative destruction through the Anglo-American hegemony: a non-Anglo-American view on publications, referees and language. Area 36:319–322

Aalbers MB, Rossi U (2007) A coming community: young geographers coping with multi-tier spaces of academic publishing across Europe. Soc Cult Geogr 8:283–302

Agnew J (1994) The territorial trap: the geographical assumptions of international relations theory. Rev Int Polit Econ 1:53–80. https://doi.org/10.1080/09692299408434268

Agnew J (2009) Globalization and Sovereignty. Rowman & Littlefield

Agnew J (2010) Still trapped in territory? Geopolitics 15:779–784

Babbitt I (1924) Democracy and leadership. Houghton Mifflin

Bialasiewicz L (2009) Europe as/at the border: Trieste and the meaning of Europe. Soc Cult Geogr 10:319–336

Brenner N (2004) Urban governance and the production of new state spaces in Western Europe, 1960–2000. Rev Int Polit Econ 11:447–488

Burke E (2008) Reflections on the revolution in France. Yale University Press

Cowen D, Smith N (2009) After geopolitics? From the geopolitical social to geoeconomics. Antipode 41:22–48

Environment and Planning D: society and Space (2003) Guest editorials, 21:131–168

European Urban and Regional Studies (2004) Euro commentaries themed section 11:335–381.

Gawlewicz A (2016) Language and translation strategies in researching migrant experience of difference from the position of migrant researcher. Qual Res 16:27–42

Geoforum (2004), The spaces of critical geography, 35, 523–558.

Gikandi S (2001) Globalization and the claims of postcoloniality. South Atl Q 100:627–658

Giraldi N (2016) Storia di Trieste: dalle origini ai giorni nostri. Biblioteca dell'immagine

Gould P, Pitts FR (2002) Geographical voices: fourteen autobiographical essays. Syracuse University Press

Gowan P (1999) The global gamble: Washington's Faustian bid for world dominance. Verso

Gutiérrez J, Lopez-Nieva P (2001) Are international journals of human geography really international? Prog Hum Geogr 25:53–69

Ivanova MN (2010) Hegemony and seigniorage: the planned spontaneity of the US current account deficit. Int J Political Econ 39(1): 93–130

Jazeel T (2016) Between area and discipline: progress, knowledge production and the geographies of geography. Prog Hum Geogr 40:649–667

Jennings C (2017) Flashpoint Trieste: the first battle of the cold war. Bloomsbury Publishing

Jessop B (2001) Institutional re (turns) and the strategic-relational approach. Environ Plan A 33:1213–1236

Jöns H (2018) The international transfer of human geographical knowledge in the context of shifting academic hegemonies. Geogr Z 106:27–37

Kirk R (2005) The conservative mind. C. Bingley

Kitchin R, Fuller D (2003) Making the 'black box' transparent: publishing and presenting geographic knowledge. Area 35:313–315

Livingstone DN, Godlewska A (1992) The geographical tradition: episodes in the history of a contested enterprise. Blackwell Oxford

Luttwak EN (1990) From geopolitics to geo-economics: logic of conflict, grammar of commerce. Natl Interest, pp 17–23

Massey DB, Allen J, Sarre P (1999) Human geography today. Polity Press Cambridge

Mayoral FM, Curiazi R (2019) Studiando il distretto industriale ecuadoriano: un'analisi comparativa tra il modello italiano e il settore manifatturiero del cuoio di Cotacachi (Ecuador). Riv Geogr Ital

Melgaço L (2017) Thinking outside the bubble of the global north: introducing Milton Santos and "The Active Role of Geography" symposium: introducing Milton Santos and "The Active Role of Geography" organisers: Lucas Melgaço and Tim Clarke. Antipode 49:946–951

Minca C (2000) Venetian geographical praxis. Environ Plan Soc Space 18:285–289

Minca C (2019) Geografia e rivoluzione. Riv Geogr Ital

Moisio S, Paasi A (2013) Beyond state-centricity: geopolitics of changing state spaces. Geopolitics 18(2), 255–266

Moisio S (2018) Geopolitics of the knowledge-based economy. Taylor & Francis

Moisio, S. (2019) Re-thinking geoeconomics: towards a political geography of economic geographies. Geography Compass 13(10), e12466.

Newman D (1998) Geopolitics renaissant: territory, sovereignty and the world political map. Geopolitics 3:1–16

Rodríguez-Pose A (2004) On English as a vehicle to preserve geographical diversity. Prog Hum Geogr 28:1–4

Rodríguez-Pose A (2006) Is there an 'Anglo-American'domination in human geography? And, is it bad? Environ Plan A 38:603–610

Sellar C (2019) Transnationalizing bureaucracies through investment promotion: The case of Informest. Environ Plan C: Politics and Space, 37(3): 461–479

Simonsen K, Öhman J (2018) Introduction: is there a 'Nordic'human geography? In: Voices from the North. Routledge, pp 1–8

Sparke M (2018) Geoeconomics, globalisation and the limits of economic strategy in statecraft: a response to Vihma. Geopolitics 23:30–37

Storper M (1997) The regional world: territorial development in a global economy. Guilford Press

Van Riemsdijk M, Wang Q (2016) Rethinking international skilled migration. Taylor & Francis

Vihma A (2018) Geoeconomic analysis and the limits of critical geopolitics: a new engagement with Edward Luttwak. Geopolitics 23:1–21

Waley P (2009) Introducing trieste: a cosmopolitan city? Soc Cult Geogr 10:243–256

2 Geography as Industry: The Institutions, Untraded Interdependencies, and Worlds of Production Shaping Academic Work

Abstract

Inclusive analyses of geographical thought should pay greater attention to the variety of institutional and organizational contexts in which the production of geographical knowledge occurs. Geographers are embedded in universities and other organizations that shape through rewards and punishments research expectations, including, but not limited to, the topic of relevance, the expectations of productivity, and the criteria to establish what constitutes 'valid' knowledge. By thinking about geography as an industry producing innovation, this chapter applies concepts developed in economic geography—namely, worlds of production and untraded interdependencies—to analyze the context that shaped Battisti's intellectual work. The results show that Battisti worked in a system of incentives not too dissimilar from Italy's manufacturing industries: a system that rewarded incremental innovation over cutting-edge discoveries, the continuation of the work of advisors and mentors, and a strong local focus. The result is the synthetic accommodation of older and newer theoretical paradigms (especially the French landscape school, quantitative revolution, and a critique of postmodern approaches), instead of the rejection of older paradigms in favor of the new, which is the more typical approach in international geography.

Keywords

Knowledge production · Geography as industry · Economic geography · Institutions · Organizations · Untraded interdependencies · Worlds of production

2.1 Introduction: The Professional Framework of Geographical Thought

Geographic thought is a subfield in human geography in its own right, a comprehensive review of which is beyond the scope of this book. It suffices to point out that there are three main recognizable trends in this subfield: first, a tendency to focus on the Anglo-American experience, especially on the connection between geography and coloniality/postcoloniality (Livingstone and Godlewska 1992; Driver 2001; Livingstone 2019). This strand of work treats non-Anglo-American geographical traditions from the perspective of their influence on the latter, with passing references to French and German geographers (Livingstone and Godlewska 1992, p. 25). Such trend is partially offset by recent attempts to broaden the inquiry to include a pan-European focus (Dikshit 2018), but the field is still far from being fully inclusive. Second, the field consists either of more or less long-term historical accounts of the discipline (Livingstone and Godlewska 1992; Boyle et al.

C. Sellar and G. Battisti, *Geopolitical Perspectives from the Italian Border*,
Historical Geography and Geosciences, https://doi.org/10.1007/978-3-031-26044-5_2

2017; Dikshit 2018), or thematic works, highlighting various strands of research and philosophical trends across the discipline (Peet 1998; Massey et al. 1999).

Anglo-American bias notwithstanding, some of the authors mentioned above have been careful in discussing the embeddedness of geographic thought in larger trends of knowledge production, as well as the political economy geographers participated in (Driver 2001). However, the goal of inclusivity requires even more attention to the variety of institutional contexts, in the measure that they shape intellectual production: simply put, we (geographers and other social scientists) choose our topics, our methods, our intended audiences, our workflow, and evaluate our writing because we participate in professional environments that determine our incentives. In this chapter, we encourage the reader to think about our discipline reflexively, as a *job*, usually performed in universities, that in the right conditions produces *innovation* in the form of new geographical knowledge, because such way of thinking unveils some of the mechanisms underpinning the *why* and *how* of knowledge production. In short, we argue that readers who know the professional incentives and constraints within which geographers operate will understand their writings better than readers who do not. Indirectly, we also suggest that such reflexive thinking may help the cause of inclusivity, because it may help to evaluate geographers' productions from traditions different from ones' own in their own merits, rather than expecting conformity with the standards of the hegemon.

To think about geography as a job means to think about the production of geographic thought as an *industry*, which in turn means to view it as an economic activity embedded in a constantly changing social world. Such shift in understanding geographic thought—from intellectual history to an embedded economic activity leading to innovation—enables to deploy reflexively the tools and notions of economic geographers to further understand some of the spatial aspects and path dependencies of the discipline. In so doing, we do not aim at a systematic analysis of the economic/social constraints shaping geographical thought, or at an in-depth theorization similar to Moisio's work on the 'political geographies of economic geographies' (Moisio 2019). Instead, this chapter has the targeted goal to understand the constraints and incentives shaping Battisti's work in Trieste from the 1970s to the 2010s. To do so, it relies on two areas of inquiry in economic geography: the first is the debate about institutions (Fromhold-Eisebith and Eisebith 2005; Bathelt and Glückler 2014; Hitt 2016). The second is the literature of innovation, and particularly the notion of *untraded interdependencies,* developed by Michael Storper (Storper 1997; Storper and Salais 1997). Albeit not the newest research, Storper's work is crucial to understanding the features of Triestino geographers' work because, first, it tied regional path dependencies with national institutional frameworks to discuss prevailing patterns of innovation. Second, he also explicitly compared the innovation incentives in the USA with a handful of other advanced economies, including Italy's. In this book chapter, we argue that some of the features of small manufacturing firms studied by Storper and others represent broader—for a better word, cultural—sensitivities that shape also academic work in universities.

2.2 Institutions as Mutual Expectations

Geographers' work on institutions has grown considerably since the early millennium, leading to an 'institutional turn'—i.e. a recognition that economic activities are embedded in, and depending on, broader social norms and organizations (Martin 2000, p. 77). The very success of the turn, and its root in the French regulation theory—which with limited exceptions provide a thorough conceptual account of the social frameworks it claimed to analyze (ibid.)—led geographers' to a rather fuzzy use of the term 'institutions.' For the most part, scholars have either understood 'institutions' as organizations (Nelson 1993), or laws, rules, and regulations (Johnson 1997; Gertler 2010). Directly related to the lack of a widely accepted, shared meaning of the term is the tendency to 'black box' institutions, and use them as a catch-all term to justify unexplained economic or social phenomena (Bathelt and Glückler 2014, p. 340). In an attempt to add analytical rigor, Bathelt and Glückler have defined them as 'stabilizations of mutual expectations and correlated interaction' (ibid.).

Focusing on mutual expectations, Bathelt and Glückler aimed at overcoming the distinction between institutions as organizations and social and legal norms. They achieved their goal by focusing on "the roles of agents and organizations, the economic practices and relationships in which they engage, …different spatial and non-spatial scales" (p. 341). Such a micro-level approach, centered on the interactions among agents/organizations, offers a bottom-up view of institutions, which emerge as cumulative effects of multiple interactions protracted over time. Once they are established, institutions contribute to the reproduction and stabilization of patterns of interaction.

Bathelt and Glückler's relational approach has several merits. First of all, it offers a dynamic view of institutions: albeit stable, they can evolve and change over time as the pattern in the relations among the actors involved changes. Second, relational patterns are not constrained by scale: actors' interactions can occur across a variety of contexts, with rippling effects at multiple scales. For example, national economic policies may be the result of the global circulation of ideas, which are introduced by actors with specific local interests in mind, and result in new, transnational flows of people, goods, and investments (Sellar 2019). Third, it accounts for —and distinguishes between—the actions of both individuals and organizations. The latter (firms, governments, trade unions) cannot be reduced to the actions of singular individuals–who in turn perform actions within multiple networks (Bathelt and Glückler 2014, p. 343). It is precisely the cumulative effects of relations between organizations and individuals that create stable patterns of mutual expectations that are not necessarily related to the goals of any agent involved, which constitute institutions if they cannot be reduced to simple regular behaviors, organizations, or formal rules (346).

Such a view of institutions has important implications for understanding innovation or—in the case of this book—the production of geographical knowledge in and about a specific place. In a nutshell, institutions establish the basic conditions for exchanging information, which is the precondition for any kind of knowledge production. Furthermore, albeit institutions are not per se spatial, they are the result of interactions between people and organizations heavily affected by borders and boundaries—which explains the dominance of the border in the Triestino school of geography, as discussed in the previous chapter. Bathelt and Glückler's view also explains various ways in which institutions may hinder the adoption of innovation: existing institutions may be tied with specific technology—thus requiring breaking older patterns of interaction to adopt new technology; organizations tend to adopt characteristics at the time of their foundation that persists over time, even when changing conditions make them suboptimal; finally, there is no guarantee that even in stable conditions institutional arrangements will lead to optimal innovation (349).

Albeit persistent over time, institutions are not immune to change. As we saw earlier, in some cases, innovation may disrupt and recreate pre-existing institutions. More in general, the interactions among actors that provide the basis for institutions are recursive: agents repeatedly perform certain actions and reevaluate their outcomes after each interaction (350). This leads to a reflexive dynamic, in which neither the actions of the individual agent nor the larger social structures to which the agent belongs can fully account for the outcomes of actions. Following Giddens' structuration theory (1984), this duality between agency and structure explains how institutions may change: on the one hand, institutions bound individual agents within structures; on the other hand, structures may change as a result of drifts in the collective actions of individuals, in a process that oftentimes does not depend on the will of any individual agent.

Viewing institutions through structuration theory explains their changes over space as well as time. Time variability is obvious: agents interactions occur in a time sequence, which implies the possibility of drifts. Spatial variability depends on localized interactions among agents: there may be average patterns of interactions, which assume a special character in a given subset of a system. Such dynamic between general and localized interactions may occur at multiple scales: global trends may assume specific national or local features; national trends may vary locally and impact transnational issues, and so on. For example, literature on neoliberalization points out that the institutions emerging out of the global diffusion of neoliberal ideology in the 1990s–2010s varied widely, mostly adapting to preexisting national and urban systems (Brenner and Theodore 2002; Brenner et al. 2010; Hirt et al. 2013).

2.3 Rethinking 'Untraded Interdependencies' and 'Worlds of Production' Through an Institutional Lense

The role institutions play in shaping firms' innovation trajectories, the role played by various spatial scales, and the links between firms, universities, and governments have been central themes of economic geography since the 1990s. While others have already reviewed and critiqued the literature on regionalism (Hadjimichalis 2006; Jonas 2012), innovation (Zhu et al. 2019; Blakely 2020), institutionalism (Gong and Hassink 2019), the role of universities (Benneworth et al. 2017), in some cases attempting to move beyond the neoliberal frameworks that underpin so much of Anglo-American economic geography (Hassink et al. 2019), our scope is narrower. Simply put, we claim that older research by Michael Storper included a description of an Italian pattern of innovation that, reinterpreted using Bathelt and Gluckler's view of institutional change, may explain some features of the evolution of geographical thought in Trieste. Therefore, this chapter's reading of Storper's work is highly selective. In particular, we are not attempting to revive 'New Regionalism,' the school of thought he founded alongside Piore and Sabel (1984) and many others (Scott 1988; Storper and Scott 1995; Storper and Salais 1997; Becattini 2001; Tabariés 2005). Over the years, their approach attracted harsh critics, that pointed out the vagueness of the concept of 'region,' the corresponding lack of analytical rigor, and the failures resulting from an indiscriminate application of the label 'regionalism' (and the cognate term 'cluster') to an inconsistent variety of policy applications (Markusen 1996; Lovering 1999). Instead, we simply aim at unveiling some of the dynamics of Italy's regional versus national innovation system that pertain to academia.

Two of Storper's concepts are particularly useful to fulfill our aim: *untraded interdependecies* and *worlds of production* (Storper 1997; Storper and Salais 1997). Simply put, untraded interdependencies are the outcomes of interdependent choices among actors (producers, but also users and regulators) that lead to certain technological innovation, which in turn lead to *spillovers*, i.e. further innovation in similar or compatible areas because 'knowing how to do one thing is frequently consequent on how to do another' (Storper 1995, p. 204). Worlds of production are the cumulative effects of untraded interdependencies. Because interdependent choices among actors require shared interpretation, precedent, and conventions, there are limited possible configurations of action in any given situation. In short, there are few possible forms of coordination among actors engaged in production and exchange, which Storper calls 'possible worlds of production' (Storper and Salais 1997, p. 20). Among advanced capitalist economies, these forms of coordination survive if they can 'continuously reinvent, differentiate, and reconfigure products to place themselves at the technological frontier' (ibid. p. 21).

In its original formulation, Storper assumed that worlds of production are largely built on conventions shaped by national and local norms, rooted in untraded interdependencies that work best when face-to-face interactions are involved, and thus explain *regional* (aka subnational, or urban) path-dependent processes. Thus, he characterized the US, French, and Italian export industries as relative coherent worlds of production, rooted in specific regional economies within the three countries: "since new successful forms of production—different from the canonical mass production systems of the postwar period—were emerging in some regions and not in others, and since they seemed to involve both localization and regional differences and specificities (institutional, technological), it followed that there might be something fundamental that linked late twentieth-century capitalism to regionalism and regionalization" (Storper 1997, p. 3).

In retrospect, such scalar rigidity constituted a key weakness of Storper and, more in general, new regionalist approaches. In this respect, Bathelt and Gluckler's view of institutions may help to overcome Storper's scalar bias. Untraded interdependencies rely upon interdependent choices that in turn depend on actors' mutual expectations that are recursively stabilized. Where and how these expectations are stabilized, and the changes that inevitably occur after many iterations of a given action, do not necessarily depend on a single scale. Revisiting Storper's treatment of Italy's economy two and half decades after he wrote two influential works—*Worlds of Production* and *The Regional World* illustrates this point (Storper 1997, pp. 134–147; Storper and Salais 1997, pp. 149–173). At that time, Italy looked 'something like the inverse of the US case, with a strong position in design-intensive products... and precision metalworking and machining [aka the machinery used in factories]... but unimpressive... high tech exports' (Storper and Salais 1997, p. 108). Production was highly concentrated in specific medium-sized towns across North-Eastern Italy: 'Udine in chairs, Como in fabrics, Sassuolo in ceramic tiles ... Bologna and Modena in mechanical engineering... [showing] an extremely strong economic specialization... a dramatically focused geography' (ibid. p. 111). Storper built his analysis of these so-called industrial districts on the work of a variety of Italian scholars (Brusco 1982; Ottati 1994; Bagnasco and Sabel 1995; Becattini 2001; Sforzi 2002), who continued their studies, training the next generation of geographers and industrial economists, throughout the early millennium. These scholars highlighted the decline, survival, and in some cases death of Italy's regional economies. As we write this chapter, the chairs district in Udine no longer exist; other districts evolved from self-contained production systems into a mix system, partly included in European value chains, partly retaining niche production (Amighini and Rabellotti 2006); more in general, leading districts firms began to internationalize heavily, starting with outsourcing production to lower wages areas in Central and Eastern Europe or North Africa and in some cases evolving into fully internationalized districts, anchored to specific locations abroad by governmental and private service providers (Rabellotti et al. 2009; Berns et al. 2020); in other cases, 'dying' industrial districts transformed themselves through a combination of redefined local institutions and immigrant firms—such as, for example, the textile district in Prato that maintained a reduced niche production while also becoming a center of low-cost apparel manufacturers owned by Chinese immigrants (Pietrobelli and Rabellotti 2010; Lan 2015).

The story of the many scalar reconfigurations of Italy's districts suggests that untraded interdependencies and spillovers change alongside market conditions that lead district firms to establish relationships with different actors (managers, other firms, government officials, consultants, etc.) in new spaces. However, the resilience of many district firms and the recursive nature of relationships among actors suggest there is a core of relatively stable institutions underpinning Italy's world of production, capable to adapt to a variety of new conditions. Quoting Moore (1993), Storper described these institutions as a 'Southern European Latin-Catholic capitalism,' that preserves political stability at the price of more rapid growth, political institutions rooted in clientelism and exchanges of favors (Storper 1997, p. 138). In the particular conditions of North-Eastern Italy, these conditions channeled a peculiar type of innovation, rooted in an incremental modification of processes and products—in which even the production equipment is constantly updated through negotiations between producers and users. The result is a 'diffused innovative capacity' that does not lead to cutting-edge innovation as in the leading US industries, but rather to the constant refinement of existing processes through practical experimentation (Storper and Salais 1997, p. 156).

2.4 Toward an Academic World of Production

The notion of Southern European Latin-Catholic capitalism is a useful starting point to shift the analysis from innovation in Italy's small manufacturing firms to knowledge production in universities. At one level, they are profoundly different kinds of organizations: one is in the private sector, the other public; one is by definition small and nimble, the other for the most part large and bureaucratic; one produces practical knowledge aimed at products' improvement; the other theoretical; one measures success in terms of profits, the other in a complex array of stakeholders satisfaction. However, at another level, they are embedded in the same larger societies, they have complex interactions among themselves and with policymakers, which in some cases may become salient features of capitalism (Moisio 2018, pp. 86–115). At this further level of analysis, it becomes apparent that some of the institutions enabling and constraining relationships among actors in the workplace, defining desirable outcomes, and shaping judgments about performances affect multiple professional fields, especially when actors in these fields work in similar places. Simply put, a university professor and a small-scale manufacturing entrepreneur in a given city will share at least some behaviors and expectations, because they are part of the same urban and national spaces.

In particular, as we saw earlier, Storper connected the notion of Southern European Latin-Catholic capitalism with institutional configurations that favor incremental innovation, rooted in constant negotiations between producers and users. It also noted the 'thickness' of political institutions rooted in hierarchical relations and exchanges of mutual favors. While Storper's analysis focused on manufacturing, Triestino academia relied on institutions leading to a similar pattern of incremental innovation. These innovation-shaping institutions were particularly visible in the areas of recruiting, evaluation of work, and relationships between the university and the city. However, as Bathelt and Gluckler pointed out (2014), institutions change over time following changes in the actors engaged in stable mutual relations. Thus, the geopolitical upheaval Trieste experienced in the twentieth- and early twenty-first-century affected actors and thus institutions, modifying the rules of the game in intellectual production—with direct consequences on the kind of geography Battisti wrote.

Reading in-between the lines of the autobiographical Chap. 3 reveals several aspects of recruitment in the profession. The chapter makes most sense when compared with similar autobiographies collected in the anthology *Geographical Voices* (Gould and Pitts 2002), featuring prominent geographers within Anglo-American academia, some of which are at the forefront of intellectual debates about the quantitative revolution and the cultural turn, which later influenced Battisti's work. The first noticeable issue is the predominance of the spaces of Trieste in shaping his path to the profession vs. the mobility shaping the experiences of the *Voices*. For example, David Harvey's autobiography starts with a school memory: "when I was twelve years old I had the first lesson on the geography of North America... we drew a map of the eastern seaboard [where I now live]... in the dark and gloomy days of postwar Great Britain I dreamed that one day I could visit North America" (Ibid. p. 150). Similarly, Battisti's autobiography recalls his high school economic geography teacher, who taught his students at the university level, blending economic and landscape geography. When Harvey longed for travel, Battisti was pushed to integrate the various aspects of the discipline in the urban horizon of his city.

The second issue, noticeable in Chap. 3, is the strong relationship with the senior faculty member who recruited Battisti in the profession, his *maestro:* such relationships feature perhaps less prominently in the *Voices* but fit well with the patriarchal institutions outlined in Storper's discussion of Latin-Catholic capitalism. More precisely, as discussed in the previous chapter, Trieste is not properly a Latin place; rather, it marks the boundary between the Latin and Germanic cultural areas (Murphy et al. 2020). Such conditions reflected on both the training and evaluation of work in Battisti's early years. Battisti's *maestro* emphasized knowledge of German, the need to integrate all aspects of the discipline, the rootedness in tradition but also the openness of new ideas—thus encouraging young Battisti to train in the—then cutting edge—quantitative revolution. Together, these were powerful forces that shaped Battisti's early intellectual focus in an analytically rigorous, oftentimes based on modeling, integration of political, economic, and historical geographies of the Triestino region with special emphasis on its relations with the border and the Austro-Germanic/Slavic worlds (Chaps. 5–7).

A third issue, not directly visible in the chapters, but extremely important in analyzing the evaluation of work, is that Triestino's geography of the 1970s and 1980s did not use the double-blind peer review system to evaluate research. Rather, senior faculty members in the national associations evaluated work openly, during career milestones at hiring and promotions. The system included sponsorships of junior faculty by senior members of the associations, in a collective and sometimes conflictual process of evaluation of candidates. While it is not our scope to discuss the pros and cons of the peer review system, already extensively done by others (Jefferson et al. 2002), it is important to notice that it is so ubiquitous in Anglo-American academia that it is difficult to conceive any other form of knowledge production (Velterop 2015). It is also beyond the scope of this book to attempt a systematic analysis of systems of knowledge production based on alternative modes of evaluation. Instead, we limit ourselves to noticing that the system described above match closely Storper's description of the exchanges of favors that are typical of Latin-Catholic capitalism. In so doing, we reinforce the argument that academia shares with other industries a variety of cultural norms. These, in turn, affect the horizon of possibilities of what is publishable and for what kind of rewards.

It is precisely such a system of evaluation that privileged theoretical and methodological continuity and refinement over sharp turns, similar to what Storper called 'incremental innovation' (Storper and Salais 1997, p. 156). Simply put, the importance of the *maestro* prolongs mentorship far longer than the relationship between Ph.D. advisor and advisee in Anglo-American academia. Therefore, in 1970s Italy, junior faculty members had the interest to follow, and build on, a mentor's research program for longer than it would have been expected in Anglo-American geography, which in turn meant a lack of incentives for sharp theoretical turns. Once more, a comparison between *Geographical Voices* and this volume illustrates the different attitudes toward cutting-edge innovation. In *Voices,* both the autobiographies of David Harvey (pp. 149–188) and Gunnar Olsson (pp. 237–286) describe the transition between the quantitative revolution and the cultural turn in human geography. Olsson pointed out that policy applications of quantitative economic geography first raised ethical issues (p. 252). Second, a scrutiny of the methods revealed fallacies—in particular, that spatial distribution of economic activities was incorrectly used to infer and manipulate underlying social processes (ibid.). Such inconsistencies led to a need for a new type of human geography. Similarly, commenting on his friendship with Olsson, Harvey claimed that 'both of us felt that some sort of modernization or even revolution in geographical methodology was imperative' (p. 166). Then, both scholars proceeded to build alternative ways of 'doing geography,' Harvey becoming the world's leading Marxist geographer, Olsson working on the philosophical principles of geographical knowledge. While we do not claim that Harvey and Olsson are representative of Anglo-American human geography, their journeys are indicative of an environment that enable the denial of earlier approaches and constructions of new ones, i.e. of Kuhnian paradigm shifts—where a new generation of scholars first challenges an older set of methods and/or assumption and then substitutes it with new ones (Bird 2014). Incidentally, such a process of innovation is similar to Storper's description of cutting-edge innovation in California, where old technologies tend to be abandoned in favor of new ones (1997, p. 154). Instead, there nothing of such sort in this volume: Battisti didn't abandon the quantitative revolution; instead, he constantly refined a realist approach to political economy, shaped by both by modeling and even older historical geography and landscape studies to adapt it to new research questions emerging in a globalizing world.

Considering the evolution of Battisti's thinking from the point of view of institutional constraints leads to a fourth point: Bathelt and Gluckler's analysis of institutional change does apply to academia. Simply put, in the nearly fifty years of Battisti's intellectual production, the 'rules of the game,' aka the stabilized mutual expectations among actors, the untraded interdependencies, and the spillovers, governing academic work have changed. Such change was especially powerful in Trieste, where academics' urban-centered policy activism of the Cold War was forced to become international in scope when the eastward expansion of the European Union (EU) opened up new relationships with its Slovenian neighbors. In a nutshell, the successes and challenges of the EU impacted Trieste directly, and in so doing pushed Battisti's work to an increasingly international dimension. Such progression of thinking is reflected in the succession of chapters in this book. On the one hand, the focus changes from the urban to the transnational, to the European and global. On the other hand, realism, the attention to historical processes, and the use of modeling inspired by the quantitative revolution remain constant.

In a nutshell, this volume is a window onto an academic world of production that shares some institutional features in common with other industries, other worlds of production more frequently studied in economic geography. Incremental innovation, a trait Storper attributed to Italy's manufacturing, is also a feature of the academic world that shaped Battisti's writing, largely due to the methods of recruitment in the profession and evaluation of work. However, the underlying institutions shaping his work were (and still are) not static but evolved to transform both these worlds of production by transnationalizing their focus. Such shift is visible in this book: the locally focused second section is based on older notes than the broader scope third section. In a nutshell, this chapter suggests that treating the discipline of geography as an industry helps to reveal the institutional processes that shape scholars' focus, methods, and theoretical approaches.

2.5 Conclusion: The Benefits of Viewing Geography as an Industry

This book aims at contributing to geographic thought, by participating in the small but growing body of work on geographers working outside the boundaries of Anglo-American geography. This chapter suggested that viewing reflexively geography as an industry may further the cause of inclusiveness by unpacking underlying institutional factors shaping diverse intellectual productions. Such an approach has the benefit of deploying—once adapted to the specific features of academia—methods and concepts economic geographers have successfully used and refined over time. In this particular chapter, we used Bathelt and Glückler's institutionalism and Storper's untraded interdependencies and worlds of productions, because the former provides tools to understand the tension between institutional continuity and change, and because the latter has already been used to analyze industries in Italy. On the one hand, a focus on institutional permanence versus transformation is necessary to understand any industry, especially in a place so deeply affected by geopolitical change like Trieste. On the other hand, Storper developed his conceptual framework to compare a variety of industries, including manufacturers located in areas very close to Trieste. Thus, bearing in mind the specificity of academic work, his work was easily adaptable to emphasize the institutional continuities between intellectual innovation in geography and innovation in other industries.

While the authors of this book highlight the benefits of discussing intellectual production as a form of innovation, the specific conceptual tools used in this chapter work well for Trieste; they should not be applied uncritically to other intellectual traditions. Instead, the goal of this chapter is to point out that economic geography offers a variety of analytical tools, which can be selected and deployed reflexively to better understand the factors enabling, constraining, and directing intellectual work. Such seldom-explored institutional constraints are not only useful to understand 'other' geographical traditions but are a constant presence in any kind of intellectual work, including hegemonic ones. Thus, readers should not only judge the content of the following chapters in their terms—as the products of a specific academic world of production—but also reflect on their context, to better understand how the institutions they live in shape the horizon of possibilities in research and publications.

References

Amighini A, Rabellotti R (2006) How do Italian footwear industrial districts face globalization? Eur Plan Stud 14:485–502

Bagnasco A, Sabel CF (1995) Small and medium-size enterprises. Pinter Publishing

Bathelt H, Glückler J (2014) Institutional change in economic geography. Prog Hum Geogr 38:340–363

Becattini G (2001) The caterpillar and the butterfly: an exemplary case of development in the Italy of the industrial districts. Felice Le Monnier

Benneworth P, Pinheiro R, Karlsen J (2017) Strategic agency and institutional change: investigating the role of universities in regional innovation systems (RISs). Reg Stud 51:235–248

Berns JP, Gondo M, Sellar C (2020) Whole country-of-origin network development abroad. J Int Bus Stud 1–25

Bird A (2014) Thomas Kuhn. Routledge

Blakely EJ (2020) In search of an innovation economic geography. In: Towards a competitive, sustainable modern city. Edward Elgar Publishing

Boyle M, England K, Farish M et al (2017) Geography and geographers: Anglo-American human geography since 1945. AAG Rev Books 5:48–61

Brenner N, Peck J, Theodore N (2010) Variegated neoliberalization: geographies, modalities, pathways. Glob Netw 10:182–222

Brenner N, Theodore N (2002) Cities and the geographies of 'actually existing neoliberalism.' Antipode 34:349–379

Brusco S (1982) The Emilian model: productive decentralisation and social integration. Camb J Econ 6:167–184

Dikshit RD (2018) Geographical thought: a contextual history of ideas. PHI Learning Pvt. Ltd.

Driver F (2001) Geography militant: cultures of exploration and empire. Blackwell

Fromhold-Eisebith M, Eisebith G (2005) How to institutionalize innovative clusters? Comparing explicit top-down and implicit bottom-up approaches. Res Policy 34:1250–1268

Gertler MS (2010) Rules of the game: the place of institutions in regional economic change. Reg Stud 44:1–15

Giddens A (1984) The constitution of society: outline of the theory of structuration. Univ of California Press

Gong H, Hassink R (2019) Co-evolution in contemporary economic geography: towards a theoretical framework. Reg Stud 53:1344–1355

Gould P, Pitts FR (2002) Geographical voices: fourteen autobiographical essays. Syracuse University Press

Hadjimichalis C (2006) Non-economic factors in economic geography and in 'new regionalism': a sympathetic critique. Int J Urban Reg Res 30:690–704

Hassink R, Gong H, Marques P (2019) Moving beyond Anglo-American economic geography. Int J Urban Sci 23:149–169

Hirt S, Sellar C, Young C (2013) Neoliberal doctrine meets the Eastern bloc: resistance, appropriation and purification in post-socialist spaces. Eur Asia Stud 65:1243–1254

Hitt MA (2016) International strategy and institutional environments. Cross Cult Strateg Manag 23:206–215

Jefferson T, Alderson P, Wager E, Davidoff F (2002) Effects of editorial peer review: a systematic review. JAMA 287:2784–2786

Johnson BH (1997) Institutions and organizations in systems of innovation. In: Systems of innovation: technologies, institutions and organisations

Jonas AE (2012) Region and place: regionalism in question. Prog Hum Geogr 36:263–272

Lan T (2015) Industrial district and the multiplication of labour: the Chinese apparel industry in Prato, Italy. Antipode 47:158–178

Livingstone DN (2019) The geographical tradition and the challenges of geography geographised. Trans Inst Br Geogr

Livingstone DN, Godlewska A (1992) The geographical tradition: episodes in the history of a contested enterprise. Blackwell Oxford

Lovering J (1999) Theory Led by Policy. The Inadequacies of the "New Regionalism" (Illustrated from the Case of Wales). International Journal of Urban and Regional Research, 23:379–395.
Markusen A. (1996) Sticky Places in Slippery Space: A Typology of Industrial Districts. Econ Geogr, 72 (3):293–314.
Martin R (2000) Institutional approaches in economic geography. Companion Econ Geogr, pp 77–94
Massey DB, Allen J, Sarre P (1999) Human geography today. Polity Press Cambridge
Moisio S (2018) Geopolitics of the knowledge-based economy. Routledge
Moisio S (2019) Re-thinking geoeconomics: towards a political geography of economic geographies. Geogr Compass e12466
Moore B (1993) Social origins of dictatorship and democracy: lord and peasant in the making of the modern world. Beacon Press
Murphy AB, Jordan-Bychkov TG, Jordan BB (2020) The European culture area: a systematic geography. Rowman & Littlefield Publishers
Nelson RR (1993) National innovation systems: a comparative analysis. Oxford University Press on Demand
Ottati GD (1994) Cooperation and competition in the industrial district as an organization model. Eur Plan Stud 2:463–483
Peet R (1998) Modern geographical thought. Wiley
Pietrobelli C, Rabellotti R (2010) The 'Marco Polo' effect: Chinese FDI in Italy. Chatham House Programme Pap 4
Piore M, Sabel C (1984) The Second Industrial Divide. Possibilities for Prosperity. Basic Books
Rabellotti R, Carabelli A, Hirsch G (2009) Italian industrial districts on the move: where are they going? Eur Plan Stud 17:19–41
Scott A. (1988) Metropolis: From the Division of Labor to Urban Form. University of California Press
Sellar C (2019) Transnationalizing bureaucracies through investment promotion: the case of Informest. Environ Plan C Polit Space 37:461–479
Sforzi F (2002) The industrial district and the 'new' Italian economic geography. Eur Plan Stud 10:439–447
Storper M (1995) The resurgence of regional economies, ten years later: the region as a nexus of untraded interdependencies. Eur Urban Reg Stud 2:191–221
Storper M (1997) The regional world: territorial development in a global economy. Guilford Press
Storper M, Scott, AJ. (1995) The wealth of regions: market forces and policy imperatives in local and global context. Futures 27(5): 505–526.
Storper M, Salais R (1997) Worlds of production: the action frameworks of the economy. Harvard University Press
Tabariés M (2005) Les apports du GREMI à l'analyse territoriale de l'innovation ou 20 ans de recherche sur les milieux innovateurs [GREMI contributions to the territorial analysis of innovation after 20 years of research on innovative environments]. Cahiers de la Maison des Sciences Economiques.
Velterop J (2015) Peer review–issues, limitations, and future development. Sci Res
Zhu S, Jin W, He C (2019) On evolutionary economic geography: a literature review using bibliometric analysis. Eur Plan Stud 27:639–660

Part II

Triestino Geographical Thought

Becoming a Geographer in Trieste. Autobiographical Essay, Reflecting on the Nature of Geography

3

Abstract

In this essay, Prof. Battisti reflects upon how his professional career from journalism to the academic world of the former Institute of Geography at the University of Trieste in the early 1970s. He shows the mechanisms of recruitment—a targeted invitation to apply that in international academia is usually reserved for senior administrators—and the affective geography of the layout of the Institute as a powerful influence on his early ideas. He also walks the reader through the relationship with his mentors, which were in turn shaped by the latest ideas of their mentors, tracing back to the foundation of the university in 1924. Such intergenerational continuity shaped his vision of the discipline as a synthetic field encompassing the human and physical domains, methodologically rooted in historically conscious, data-driven analyses.

Keywords

Autobiography · Trieste · Mentorship · Methods · Geographical traditions

3.1 Joining the Institute of Geography

I decided to become a geographer roughly one month after I graduated from the University of Trieste. It was not a something I had pursued; instead, it was the acceptance of an unexpected offer—a moment of light that came in one of my darkest hours. That very day I had met our family doctor, who informed me that my father had a cancer, and it was incurable, so we should prepare for the worst. I remember walking to my mother's home to bear the news, finding her on the doorstep, and, before I could say anything, she uttered "The University called. They are offering you a job."

The call came from the Institute of Geography (Fig. 3.1), which I admittedly didn't know well. I took a few classes here and there as a student, got passable grades, but I sincerely did not expect the opportunity. Moreover, I already had two pending job offers—one from a large manufacturer of maritime engines, the other from *Generali*, Italy's largest insurance company. These were both attractive positions from stable employers, but at the time, I had been dreaming

C. Sellar and G. Battisti, *Geopolitical Perspectives from the Italian Border*, Historical Geography and Geosciences, https://doi.org/10.1007/978-3-031-26044-5_3

Fig. 3.1 University of Trieste main building: historical site of the Institute of Geography (1961). *Source* University of Trieste archive

of an academic career for some time already. I also knew it was a steep path, so I hadn't put much hope into it. That call left me surprised and intimidated, because I had never thought about *geography* as an opportunity, so I felt utterly unprepared.

I was especially surprised because I did not see much connection between geography and my degree in industrial and commercial economics. In retrospect, I realized I was not as unprepared as I thought, thanks to my earlier experiences in journalism. I had already benefited from such experiences by writing my thesis, a ponderous work titled *The Market and Promotion of Aerospace Products*. As a student, I had already published several newspaper and magazine articles on space exploration, which at the time—1960s and early 1970s—was the preferred topic in science journalism. I was young, ambitious, and confident of my expertise, so I was hoping to analyze the whole aerospace field—at the time limited to less than a dozen countries. Without even knowing it, I had started my journey as an economic geographer by looking at the regional and national specificities of the global industries involved in the space race. By the time I joined the university, I had published enough newspaper and magazine articles to enroll in the Order of Journalists—which I still proudly keep. That kind of writing taught me to constantly collect data, interpret them, and write succinctly, all helpful skills in my transition to academic work. I was already capable of the 'geographical synthesis,' which is a crucial feature of good writing in the discipline.

At the time, I wasn't fully aware of how my various skills and experiences fit together; so I asked the advice of my beloved high school literature and history teacher, Mr. Glauco Dorsa. I told him "I have been offered three jobs: you know me well, please tell me what I should do." As I expected, his reply was sharp and synthetic: after reviewing the pros and cons of corporate work, he told me "university careers are extremely slow, you will have to tighten your belt for a long time, but in the end, the university is the only place that will allow you true freedom." He was right.

I remember vividly the long hours I spent behind an old wooden desk my boss ordered out of storage for me. They put the desk in the middle of a long corridor; on the walls, in front and behind me there were rows and rows of books and journal collections, in all major Western languages, dedicated on the widest possible range of topics and subfields in geography. It was the heart of the Institute, and, in the early 1970s, Italy's best academic library in geography. Years before the Internet, it had subscriptions to the most important international journals, and publications exchange agreements with a large number of universities, several of which in Germany. By the time I joined, the Institute's director, Professor Eliseo Bonetti, had personally curated it with passion and competence for roughly thirty years. Without ever moving from Trieste, he had been able to follow worldwide trends, keeping the Institute up to date with the cutting edge of the discipline in the 1950s–70s.[1] In so doing, he had built the perfect tool to open my youthful mind to new knowledge.

3.2 Understanding Triestino Geography

The Institute was also an ideologically and epistemologically 'neutral' place: the director embodied tradition and innovation in equal measure, which he both shared harmoniously with selected few students. The intimacy of learning grew with its diminishing size: between the 1930s and '50s, it had been one of the larger units on campus; since then it shrank because several faculty members had left it to start new academic units. For example, one scholar founded a Geography Lab in the School of Letters and Philosophy; another started a parallel Institute within the School of Education; many more were employed as standalone geographers in a variety of departments and schools, including political science, foreign languages, and the School for Interpreters and Translators. For me, all of this meant I could benefit from both a high level of attention from my mentor and at the same time thrive in an expanding field, all within the close proximity of the same urban space.

Because I knew so little, I tried to read as much as I could, to get a sense of the discipline, as well as to start my own research program. I remember the feelings of dismay I felt every time I encountered a new topic and a new approach in the discipline, sensing they may never end.[2] So early after graduation and under Prof. Bonetti's guidance, I felt I had to read everything before trying to produce original work. Such feeling lasted until my own nature prevailed and came to the obvious conclusion: there was only so much I could read, even considering the many decades I was hoping to spend with the Institute. I accepted I would have never mastered it all; instead, I could learn quite a lot about a few specific places and topics. Without even knowing it, I had finally grasped geography.

Beujeau-Garnier wrote "geography is like the lobby of a train station, everybody meets there" (1963). Within the discipline there is even space for the work of non-geographers (Corna Pellegrini 2007); so I knew I was going to find my place, I just needed to work hard and wait until I gained a bit of confidence in my abilities. Today I am even more convinced that I didn't land at the Institute by chance or blind faith: it was, borrowing the term from Ortega y Gasset, a *vocation*.[3]

As I am writing this piece, fifty years have passed since I was hired; it still feels like a miracle.[4] Both my austere mentor and the

[1] For example, he introduced in Italy Walter Christaller's location theory (Battisti 2006).

[2] Later I understood the value of the eclectic frame of mind I had learned as a journalist. Early in my academic career, I rejected it, due to the enthusiasm of the neophyte, which made me embrace the notion of specialized knowledge. Such attitude was the legacy of undergraduate education, at the time based on rigidly distinct disciplinary fields. Later I learned that academic innovation requires cross-disciplinary fertilization to tackle new ideas and topics. Moreover, in geography, more than in other disciplines, specialization is not limited to the object of analysis but includes the intersection between the area of inquiry and the chosen methodology.

[3] Vocation is a central theme in Ortega y Gasset's philosophy. Gaete (2016) synthesized it as follows: El hombre solo aprende lo que él es, siéndolo. Ensaya multiples caminos. Y cuando encuentra el bueno, experimenta "el prodigioso fenómeno de la felicidad" [man learns who he is only by being. He tries multiple paths. When he finds the good one, he experiences "the wondrous phenomenon of happiness"].

[4] My hiring—especially the fact I received a call out of the blue—is unheard of in Italian, and I guess any other academic job. I learned only later what happened. Roughly translating Italian administrative structures in a language familiar to an Anglo-American audience, the

colleague who gave me the call made clear that my fortunate hire was not the end of it, and that tenure required deep, intense, and continuous work that eventually shaped my whole intellectual life.[5] Such work ethics is the core of a profession unlike any other. Like many Anglo-American graduate students working as teaching or research assistants, I joined to work for my mentor. However, assisting a senior faculty is only the first step toward 'scientific maturity' that should lead to full professorship. At the time, in Italy, there were precious few full professorships available, and there were no stable intermediate steps. The ranking of associate professor was established only in 1982. When I was hired the law had just granted the possibility of tenure (without automatic promotion) to assistants, who became roughly equivalent to assistant professors in the USA. It took me six years of patient and continuous growth to get tenure, and many more to get promoted.

In my early days at work, my mentor told me "See, Dr. Battisti, geography is a difficult subject. First, we have to collect cartography, then study the physical environment, then the historical evolution of territory, the characteristics of its population, the written and material culture, the economic assets, and the political conditions. Only after all of this is done we can start the actual scientific work." It was a necessary introduction because my university degree was not in geography, but how heavy to digest! It was also in line with the severe character of my mentor, who had welcomed me on my first day at work asking "do you speak German?" Modern geography had started in Germany; moreover, Professor Bonetti was a student when Trieste was still part of the Austrian Empire, so it was a natural question for him to ask. My positive answer—which he did not take for granted because I had not taken German language classes at the university—reassured him.

Even though he held the title 'Professor of Economic Geography,' he felt very strongly that geography ought to be a synthetic discipline, encompassing the physical and human domains. He kept his firm belief, rooted in 19th-century landscape studies, all his life. At the same time, he was also open to innovation, so he allowed me (first in the Institute) not to train in the traditional studies of landscapes and rural housing to focus on what at the time was called *new geography*—now known as the quantitative revolution—characterized by deductive reasoning, modeling, and the use of spatial statistics. In the early 1970s, it was the dominant paradigm in international geography, just starting to influence Italian geographers. It also carried a vision of the world I found fascinating. I was indeed in the right place: Trieste's Institute of Geography was at the cutting edge of this process of renewal. Much later, my colleague Angelo Turco—who due to another twist of faith got one of the full professorship positions that had been offered to me—remembered fondly Professor Bonetti as a pioneer whose work opened up the closed horizons of Italian geography (Turco 2010). Indeed, he

Institute had asked and obtained a new tenure-track line in economic geography to hire a certain Marzio Strassoldo. He had just graduated with a highly innovative thesis, the first in Italian academia to use Christaller's location theory; he had also shown dexterity with quantitative methods. Amid the quantitative revolution in geography, he was a very appealing candidate (Strassoldo 1973). Unfortunately for geographers, the larger and better-funded Institute of Statistics offered him a job first. Therefore, to avoid losing the tenure-track line, geographers had to find a suitable candidate fast. They didn't have anyone in the pipeline, so they selected among recent graduates in economics, looking at students with the highest GPA, and paying particular attention to the grades in math, economics, statistics, and foreign languages—which were considered the most relevant areas of expertise for an economic geographer. The selection committee consisted of the Institute director and a junior faculty member, Dr. Maria Paola Pagnini, who now leads the doctoral program in political geography at Nicolo Cusano University in Rome.

[5] Three years later, my mentor, Dr. Bonetti started the meeting of the tenure committee by stating "Let it be clear that I am willing to fail anyone, including my own candidate." Unlike in the USA, tenure and promotion committee meetings may be tough and confrontational, with mixed commissions of internal and external senior faculty members who sometimes grind the candidate, who has to defend his or her research accomplishments. For most candidates, the mentor's presence helps, but I have no doubt Bonetti would have failed—and thus fired—me if I had not convinced him of the quality of my work. Italian faculty members are recruited by cooptation, but we must adhere to rigorous ethical principles when called to evaluate candidates.

had been Italy's first full professor to endorse the quantitative revolution.

While writing my story, there is another invaluable mentor I feel obliged to remember, my high school economic geography teacher, Mr. Alessandro Cucagna (Battisti 1998). He taught in beautiful and perfectly delivered lectures, which doubled the content of his courses because he interspersed the government-mandated curriculum with a synthesis of the human geography course he taught as an adjunct at the university. Even though my classmates and I were trained stenographers—we even won a national competition—we could not write down all he had to say. Mr. Cucagna was a polymath, an excellent cartographer, and a teacher that strove to convey the relationships among all that is visible on the planet, in the soil, the subsoil, the seas, and the atmosphere. It was way too much; needless to say, many of us failed his classes. However, we all agreed his lessons were extraordinary—including those among us who failed and those who, 55 and more years later, still feel anxious just by mentioning his name.

My elementary and middle school teachers taught me a dry, boring version of geography. Mr. Cucagna changed my mind; his evidence-based lectures conveyed a version of the field that was exciting and engaging because he aimed at a synthesis of knowledge. The direct outcome of classes was I realized I needed to pursue intellectual work. Bookkeeping and accounting—the profession my high school was training me to pursue—ceased to be my goal. Thus, I decided to try science journalism, starting the journey that eventually landed me in academia.[6] Two of the best Triestino geographers shaped my early life. Much later, I encountered again Mr. Cucagna: he had left high school to become a full-time university professor.[7]

[6] Then I went to college. University studies—even though journalism took the bulk of my time—deeply affected me. I soon realized that life in a newspaper's editorial staff was too narrow for me. So, halfway through university, I refused a job offer in the city's paper, hoping in something better.

[7] After his retirement, I inherited his professorship, his desk, and even the position of chair of the Department of Historical and Geographical Sciences, which was the evolution of the Institute of Geography.

3.3 Developing My Vision of the Discipline

Both my mentors taught me a synthetic vision of geography, both human and physical, but honestly, so early in my career, I could not reflect much on the nature of the discipline. During my early years, I found so many gaps in my preparation, I felt so much pressure to publish as soon as possible, so much preoccupation to produce good quality research that I didn't have time to think too much about epistemology. In those years, in Italy as well as in Anglo-American academia, physical, and human geography began to split, with physical geographers converging in the more promising field of earth sciences. Moreover, science journalism notwithstanding, my knowledge of natural sciences was weak. Even on the human side, I had to learn a lot about economics and economic history. However, the experience of my unfortunate first attempt at full professorship—another colleague got the position prepared for me, but a parallel commission declared I was qualified also for senior positions in human geography—broadened my horizons and boosted my confidence. Only later, thanks to the collaboration with my first student, prof. Andrea Favretto,[8] I published some work in physical geography. In short, I took a path opposite to my mentors' wisdom: they felt a geographer should start with earth science to progress to ecology and culminate in human geography. I started with economic geography and later branched out to human geography, which deals also with earth science. I don't know everything in geography, but I still attempted to practice an integrative discipline, which nowadays, in an ever-fragmenting discipline, seems to be a lost art.

Readers—those who had enough patience to come this far—can now understand my vision of the discipline and my method. It's a simple way of working, albeit not always easy. I start by following the historical events that built the

[8] Thanks to his expertise in IT, he is now Italy's most prominent expert in remote sensing and automated cartography.

'geographical facts' I investigate. I look at human and natural history together, at tangible features and ideas, especially the projects and visions that underpin human actions. In a nutshell, 'my' geography is a historical geography of specific places.

Although quite a few colleagues share similar views,[9] there are some nuances in my work. I am convinced that geography is at the same time a humanistic and scientific discipline. Humanistic because humanity remains the focus of the discipline, especially concerning the spatial dimensions of social groups. Scientific because it deals with measurable phenomena, part of a reality that our senses register as concrete and tangible, and therefore 'real.' In a nutshell, I am a realist in an age in which many geographers prefer to focus on cultural interpretations, on 'imagined geographies'—the multiplication of which seems to be endless.[10] I have no quarrels with the work of physicists that emphasize the dual nature of matter as both energy and particles. They are in a different field, one that works at scales incomparably different from geographers, so different to be irrelevant to our work.

Scale, a defining concept in geography, is also a crucial factor to differentiate among academic disciplines because an understanding of scale contributes to an effective conceptual fragmentation of reality; the latter must be an effective tool of work, not an unpassable barrier. Therefore, the various subdisciplines—physical, biological, anthropic, economic, theoretical, applied, regional and general—should all be parts of a unified geography. A geography whose core precept is to unify the human and natural worlds. Such unification is sorely needed in these years of environmental crisis, to respond to the—mistaken—dominant view that privileges the ecology of plants and animals at the expense of humans and their necessities.

The synthetic nature of the discipline should be understood as the collective body of work of its practitioners, geographers who must focus out of necessity and limits of human nature on narrow topics. The human condition allows reductionism, to later bounce back to increasingly complex syntheses, without ever hoping to reach absolute knowledge, which is tantamount to closing the book of life. Such an aim transcends human capacity. Readers can understand such drive toward synthesis by looking at Carl Ritter's work (1779–1859). He passed away before completing his description of the world, consisting of 19 volumes, which barely covered two continents. His student Elisée Reclus (1830–1905) from his side wrote also 19 volumes but was able to complete Ritter's universal geography project. Those were different times and a different science; today's modern and postmodern geographers have to work in increasingly narrow spatial, temporal, and methodological frames. This is not an excuse to stop striving toward a unified body of geographical knowledge.

[9] I am particularly sympathetic to the "oriented process" approach in regional geography, championed in Italy by Vallega (1995). Vallega was a tireless worker, a true gentleman, and a friend of our Institute of Geography. He was president of the International Geographical Union, until his untimely death due to illness.

[10] I believe the warning that Francesco Compagna gave leftist geographers during the upheavals of 1968 is even more relevant today: "my children, you cannot get rid of [concrete] places, otherwise you will become castaways in an imagined geography" (1970, p. 900).

References

Battisti G (1998) Alessando Cucagna (1917–1987). Boll Soc Geogr It 11(5):471–473

Battisti G (2006) Eliseo Bonetti: ritratto di un geografo [Eliseo Bonetti, portrait of a geographer]. Riv Geog It 63(n/a):367–369

Beujeau-Garnier J, Chabot G (1963) Traité de Géographie urbaine [Treaty of urban geography]. A. Colin, Paris

Compagna F (1970) Figlioli miei. In:La ricerca geografica in Italia 1960–1980 [Geographical research in Italy, 1960–1980]. Corna Pellegrini G, Brusa C,AGEI Varese, pp 877–900

Corna Pellegrini G (2007) Geografia preziosa e diversa. Il pensiero geografico in altri pensieri umani [Precious and diverse geography. Geographical thought in other human thoughts]. CarocciRoma

Gaete A (2016) Vocacion y autenticidad [vocation and autenticity]. Revista De Filosofía 8(2–3):85–112

Strassoldo M (1973) Le località centrali nella Bassa Friulana [Central places in Friuli's low plain] Camera di Commercio Industria Agricoltura e Artigianato, Udine

Turco A (2010) Configurazioni della territorialità [Configurations of territoriality]. F. Angeli, Milano
Vallega A (1995) La regione. Sistema territoriale sostenibile. Compendio di geografia regionale [The region. Sustainable territorial system. Compendium of regional geography]. Mursia, Milano

4 Living in the Borderlands: Political Geography, Geopolitics, and Advocacy in the *Triestino* School of Geography During the Long Twentieth Century

Abstract

Trieste is the seat of a distinctive school of geographical thought that developed since the foundation of the university in 1924. Like scholars in other disciplines, policy activism characterized Triestino geographers, driven by the geopolitical upheavals of the twentieth century. Italy took over the city from Austria after WW1 and nearly lost it to Yugoslavia after WW2. These events threatened the socioeconomic fabric of the city and mobilized local academics in its defense. Post-WW1 reconstruction and post-1929 protectionism challenged the livelihood of the port, leading geographers trained in the French landscape studies to focus on both political and economic geography, and later to found the first Italian geopolitical journal. After WW2, policymakers tasked Triestino political geographers with developing new approaches to study the complex, dynamic realities of the Iron Curtain, as well as re-establishing relations with Yugoslav academics. Finally, the post-Cold War environment led them to focus on the new territorialities of an enlarged European Union. In a nutshell, Triestino scholars have been experimenting with approaches and methods in political geography and geopolitics because the grand geopolitics of global conflicts and upheavals had such a direct, sweeping impact on their city and region.

Keywords

Cold War · European Union enlargement · WW1 · WW2 · Italy · Yugoslavia

4.1 Introduction: The Civic Commitment of *Triestino* Intellectuals

My whole career and the careers of all my mentors are embedded in the century-long *Triestino* School of Geography, an intellectual tradition with a character of its own, at the same time determined by, and actively shaping the history of Trieste through policy advocacy. Faced with the terrible tragedy of two world wars ravaging a city that had just reached the peak of its development, a group of scholars, working at the newly founded University of Trieste (est. 1924), took up the challenge of the events, trying to analyze the transformations of reality that the pendulum of history imposed from time to time. The method was always scientific, but the purpose to which it was addressed was based on a sense of civic duty, that is, the duty incumbent on intellectuals to contribute to the solution of society's problems.

In the microcosm of Trieste during the 'long twentieth-century,' society had to come to terms with a continuous subversion of the preexisting

C. Sellar and G. Battisti, *Geopolitical Perspectives from the Italian Border*, Historical Geography and Geosciences, https://doi.org/10.1007/978-3-031-26044-5_4

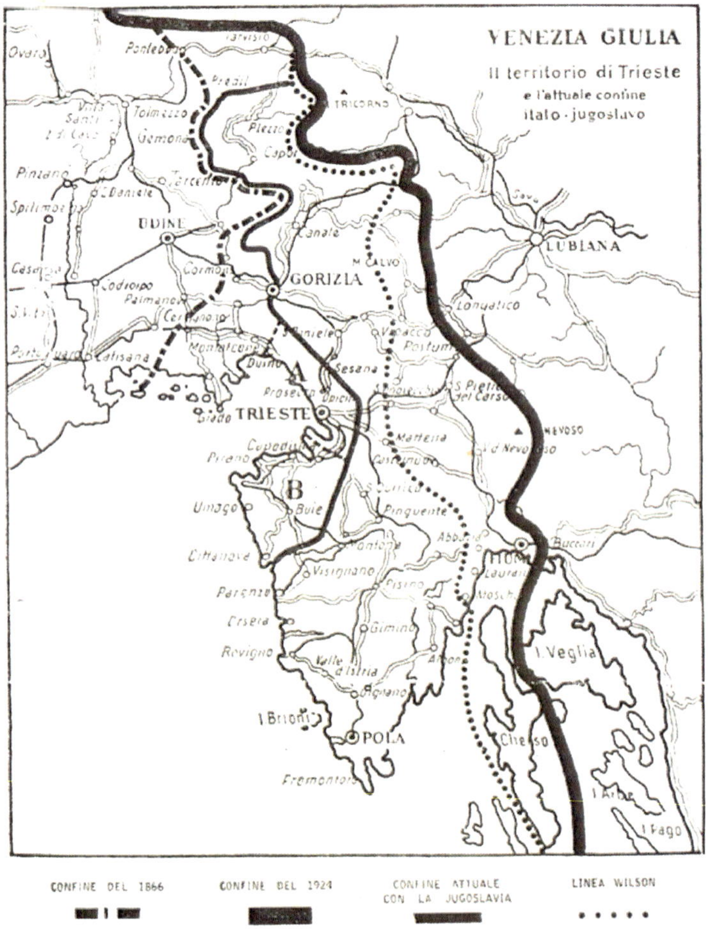

Fig. 4.1 Two centuries of borders in the upper Adriatic area. *Source* Barbi (1975)

geopolitical order that led to the border shifts shown in Fig. 4.1. Border shifts of such magnitude carried inevitable repercussions on the terrain of the spatial organization of the economy. During these years of upheaval, geographers were not alone in deploying their intellectual skills to protect the city. Legal scholars attempted to tackle the juridical problems stemming from the disappearance from one day to the next of the Austro-Hungarian Empire, working at the leading edge of international law. M. Udina led the group of jurists committed to the service of the community, which included E. Cammarata, *Rector Magnificus* (aka, the university chancellor) who defended the national choice of the city after WW2, and many others (Udina 1969; Gerin 1952, 1953, 1973; Bevilacqua 1984).

Historians worked alongside the legal scholars, and already at the end of the 1930s, Schiffrer wrote *Le origini dell'irredentismo triestino (1813–1869)* [The origins of the Triestine irredentist movement, 1813–1869] (1978), the first critical analysis of the events leading to the separation of the city from Austria. Schiffrer's intellectual production was too rich and complex to remain within the confines of the historical discipline alone. He worked at the Institute of Geography of the university during the difficult years of the Anglo-American administration, alongside Roletto, Bonetti, and other collaborators of the Institute.[1] His *Saggio di una carta dei limiti nazionali italo-jugoslavi* [Sketch of a map of the Italian-Yugoslav national boundaries] (1946) written in a climate of terror, under the threat of ethnic cleansing, represented the most important scientific contribution made by Italy at the Peace Conference. Other civic-minded historians include Cusin (1937), a precursor of studies on the origin of Italy's eastern border, Maserati (1963), Marsico (1979), and others, up to a newer generation writing at the end of the century (Pupo 1979; Valdevit 1984).

After World War II, novelists and storytellers joined academics in the defense of the cultural and political legacies of the city. Quarantotti-Gambini (1951) wrote an accurate and emotional description of the occupation of Trieste by Yugoslav forces in May 1945, while Tomizza (1976) unveiled the harsh and simple world of the inner Istrian peninsula, crossed by the tragedies of two world wars. These are only two of the many writers that the scholar and novelist Magris (1980), Ara and Magris (1982) will analyze to define an ambiguous, difficult to capture cultural reality, probably best defined as 'frontier.' Besides Magris, many tried to decipher the "myth" of Trieste (Lokar 1971), including journalists such as Ronchey (1973), and Bettiza, who wrote reports on the Eastern border (1966). These scholars, novelists, and journalists understood that frontiers and borderlands all resemble each other, once the mechanisms that characterize them are unveiled. Thus, they expanded their analyses beyond the city to include Central Europe, analytically recasting the notion of *Mitteleuropa*, understood as a place of collective memory and at the same time a dimension of the spirit (Agnelli 1971; Magris 1980).

While these public intellectuals moved in the foreground, businessmen and diplomats kept working in the region, albeit more discreetly. At the end of the 1960s, The Italian government founded the Institute of Studies and Documentation on Eastern Europe (ISDEE 1989) as a part of a joint diplomatic effort with Yugoslavia, which later transformed into the Institute of Studies and Documentation on the European Community and Eastern Europe. Up to its closure in 2010, ISDEE carried out a full program of meetings and study seminars aimed at promoting collaborations between Italian scholars and colleagues from Yugoslavia and its successor states and the rest of Southeastern and Central Europe. ISDEE also produced documents and research to support public and private agencies, and published its own journal, *Est-Ovest*, targeting both an academic and practitioner audience in Russian and Eastern European studies. In the same years, the Institute of International Sociology of Gorizia (ISIG 1990) was founded by sociologists working at the Universities of Trento and Trieste. The Institute initially activated a fruitful interdisciplinary collaboration aimed at the "discovery" of the border, including economists (Sambri 1970) and geographers (Valussi 1971) alongside sociologists. In this context, fraught with difficulties and therefore rich in stimuli, the work of the geographers from Trieste should be framed.

4.2 The Interwar Period: The Formative years of *Triestino* Geography

With the arrival of Roletto in 1927 and the foundation of the Institute of Geography, *Triestino* academics began a research program along the consolidated lines of the geography of the time. Early studies concerned the mountain areas, where it was easier to trace the outcomes of the encounter between man and nature, the focus of the then-dominant paradigm of 'landscape studies,' developed by the French geographer Vidal De La Blache (Roletto 1915). Border issues were an equally important theme of research in Italian geography because of the aftermath of WW1 (Adami 1931; Migliorini 1932); due to Trieste's role as a borderland, local scholars contributed actively (Cumin 1933). However, the most serious problems that

[1] After Italy lost WW2, from 1946 to 1954 a joint US-UK government ruled Trieste to prevent further conflicts between Italy and Yugoslavia, which could have triggered a larger confrontation between the East and the West.

troubled the world in the interwar period were economic: reconstruction, the "great crisis" that followed the market crash of 1929, the rollback of the Gilded Age free-market policies in favor of protectionist controls of markets and raw materials. Geographers were aware that the solution to these problems required political action: political geography, therefore, went hand in hand with geoeconomic analysis.

Professor Roletto became soon involved in political and economic analyses by necessity because he led an institute in a place such as Trieste, which had to reconstruct its economy and its regional footprint at the same time due to the border shifts after the war (Roletto 1932). After a meeting with a younger scholar, E. Massi, he identified political geography as a field of study particularly suitable for dealing with emerging issues of his time on a scientific level (Valussi 1965). Such a decision brought Trieste to the cutting edge of Italian geography because this branch of the discipline had just become autonomous with the treatise of De Marchi (1929), published alongside a similar book on economic geography by the same author for the Bocconi of Milan.

De Marchi's work shows the effort to adapt to the needs of the new times, produced by an authoritative exponent of physical geography. The assumption underlying his work was that Geography was a single discipline, articulated in the human and physical dimension, which soon sparkle a lively debate. A decade later, Jaja (1939) challenged the definitions of De Marchi, which he considered excessively deterministic: "Political geography is the science that studies the influence of natural factors and their distribution on the earth's surface in the formation, development, and decadence of states. His vision, which stems from the integral conception of the discipline, wants Political Geography to be "the Geography of States (the distribution and the purpose are implicit in the name of Geography) … the study of the influences of the natural environment and the human environment associated. *With this to add, the influence of the natural factor, if it is right that it has to precede in order of time, or it is a point of departure [for the formation of states], it nevertheless matters the least in the formation, or development, or decadence of the states*" (the italics are mine). On the one hand, Jaja rejected De Marchi's position on the direct influence of nature on politics; on the other, he maintained strong conceptual links between physical and human geographies: "Given that physical geography is the study of the mutual interactions between natural factors, political geography is the physical geography of states, economic geography **is** physical geography of the economic life of peoples, especially states."

The growth of political geography as an autonomous subdiscipline was a crucial part of the gradual shift of emphasis of our discipline from physical to human-centered studies which, in the 1930s, culminated with the development of geopolitics. The adoption and popularization by the Nazis of notions such as 'vital space,' originally developed in biogeography and then adapted to human-centered analyses in geopolitics, became part of a dangerous challenge to the laws of political and economic balance worldwide. Fascist Italy adopted such narratives in its foreign policy, with disastrous results culminating in the loss of WW2. While governments borrowed notions from geopolitics to justify aggression, academics kept developing it because it seemed to offer, between the thirties and forties, a scientific approach to analyze the foreign policy gambles that many governments were playing.

In Trieste, thanks to the resourcefulness of Massi,[2] Roletto became in 1939 the editor of the first Italian journal of geopolitics, (*Geopolitica. Rassegna mensile di geografia politica, economica, sociale, coloniale [geopolitics. Monthly review of political, economic, social, and colonial geography]*, building on the work of German scholars to develop an Italian school of

[2] After the war, Massi will see his academic career interrupted due to his political position. Once moved to Rome, he will take over the leadership of the local geoeconomic school. Eventually, he will be elected president of the Italian Geographical Society, a position he held from 1978 to 1987 (Massi 1986).

geopolitics.[3] However, Massi and Roletto did not resolve the epistemological uncertainties surrounding the distinction between political geography and geopolitics. In fact, in the draft of their treatise on political geography written by the two authors (Roletto and Massi 1931), next to De Marchi's definition of political geography we can still read, annotated in Roletto's handwriting, this second thought: "would [geopolitics] be: the science that studies and evaluates singularly and as a whole the geographical coefficients of the life of states, or: [the science] that studies and evaluates the geographical factors as essential coefficients of the life of the state"?

The two scholars' biggest ambition was to uncover trends in the evolution of states. They specified they aimed at understanding "trends and not laws," because in addition to geographical factors, important and uncontrollable factors of political, moral, and sentimental order exert their influence. Such an evolutionary approach led them to draw a distinction between Political Geography and Geopolitics, the first seen as a branch that studies the states in their static position on the earth's surface, the other as a branch of politics that considers states as territorial organisms in their dynamics, that is in their movements and their developments (Roletto and Massi 1931). It followed that the essential difference between political geography and geopolitics would be the strictly analytical nature of the former and the predictive nature of the latter. However, the two authors never committed to understanding geopolitics as a discipline separated from political geography, because they saw a danger in separating the dynamic and analytical aspects of political geography.

In the end, Triestino scholars of the interwar period failed to develop geopolitics in a full-fledge scientific endeavor. Its epistemological ambiguities could not be solved with the methods available at the time. It took the introduction of statistical tools in political science in the late twentieth and early twenty-first centuries to achieve reliable analyses of political trends. Epistemological weaknesses failed interwar geopolitics as much as its use by the axis powers to justify their actions during WW2, including that of Trieste, so much more balanced than the German discipline built by Haushofer (1925, 1927). Such failure was already implicit in the handouts of political geography edited by Roletto and Massi, who succeeded De Marchi as leading professors in the Faculty of Political Science: "the term geopolitics, which we will often use in the course of the lessons should be understood as a term corresponding to that of dynamic political geography. So in the term itself will be—after what has been said—incorporated the static part [of the discipline]" (Roletto and Massi 1931).

4.3 After WW2: Geographers' Policy Activism in a Bipolar World

The tragedy of WW2 discredited not only geopolitics but also political geography. In the 1950s and 1960s, Triestino geographers abandoned it to pursue other endeavors, especially after the allied powers returned Trieste to Italy in 1954. Bonetti (1960) turned to historical geography to analyze language communities in the Fella River basin, while Cucagna (1958) focused on the administrative boundaries of the mountain communities, and Lago (1966) studied the mountain town of Tarvisio as the center of a transborder region. Only later, the anthology of Kasperson and Minghi (1969), an authoritative, international-level publication re-started *Triestino's* interest in political geography, because Minghi, who soon thereafter joined the faculty at the University of South Carolina, had strong connections with Trieste. It was Minghi's influence that lead Maria Paola Pagnini, Bonetti's first student, to frame her work within political geography, which later led her to a professorship in the Department of Political Science (1987, 2001, 2011, 2015).

[3] While not disdaining political commitments—According to Bonetti, he even aspired to be appointed governor of the Free Territory of Trieste, established by the Allied powers in 1945, hoping it could become a permanent reality—he refused the offer to transform the journal into an instrument of the regime.

In the 1960s and 70s, however, the most prominent political geographer was Giorgio Valussi, who studied the new reality that had arisen on Italy's eastern border (1957, 1971, 1972). It was a challenging task because he was still working with traditional, rather static, approaches that were inadequate to analyze the complex, ideology-laden processes occurring at the border. More dynamic approaches emerged later, in the 1980s (Turco 1984), but Valussi, although aware of these new theoretical developments, did not feel prepared to adopt them. He was an "organic" geographer, trained in French landscape studies and Ratzelian geography in the wake of Roletto. Similar to Roletto, Valussi engaged the policy and political world of Trieste and the neighboring region of *Friuli*. On the one hand, policymakers asked him to study the new regional reality created by the territorial losses of WW2 and the reunion of Trieste with Italy in 1954. Such policy-relevant work earned him the role of "official geographer" of *Friuli- Venezia Giulia*, the new administrative unit born out of the merger of *Friuli* and Gorizia with the territory of Trieste. On the other hand, his work was instrumental to Italy's foreign policy, especially concerning the various initiatives that multiple levels of the Italian government were pursuing to improve the relations and to re-open the border between Italy and Yugoslavia.

Valussi's policy activism was met with misunderstandings and hostility among *Triestino* academics.[4] In the 1960s and 1970s, an attitude of closure toward the outside world still permeated Trieste, due to the multiple traumas experienced after the war—such as the loss of the hinterland, the separation from the functional region of the city, the amputation of the town territory, and the persistent uncertainty on the future. After promotion to full professor, Valussi left the straitjacket of Trieste to assume a pioneering role in the contact with the young Slovenian geography through the organization of several symposia with his colleagues across the border. After the "political earthquake" that took place in Trieste after Italy and Yugoslavia finally signed a peace treaty in 1977—the Treaty of Osimo (Cecovini 1977)[5]—Slovenian geographers too began regarding Valussi with diffidence, perhaps because they were alarmed by some balanced stances in favor of the now tiny Italian minority in Slovenia and Croatia (1976). These fears were unfounded, as he was deeply alien to political partisanship, preferring a rigorous, facts-based approach even when dealing with highly controversial policies.

The chronological sequence of Valussi's studies allows us to trace the evolution of the situation on the eastern border, whose volatility and ambiguity are shown in Fig 4.1, representing the situation after WW2 and before the official peace treaty of Osimo in 1976. The timid opening of the borders after the agreements that led to the reunification of Trieste with Italy in 1954 allowed the emergence of local cross-border tourism, based on family relationships. The limited financial means of these tourists did not balance the profound distortion of the flow of goods caused by the split between Eastern and Western Europe (Valussi 1957). In front of a conflict extended to a planetary level, it seems indeed comforting to revisit the quarrels that once opposed the small prealpine communities (1962). From the interaction of these problematic nodes configured at different scales emerges the "personality" of Friuli-Venezia Giulia, as a borderland between Italy and the Danubian world (1962).

1976 was the crucial year in defining the character of *Triestino* geography as a discipline

[4] Strong-will man and tireless worker, Valussi gained a preeminent position among Italian geographers. Alongside the research activity, he transformed the Association of Italian Geography Teachers (AIIG) into the most important club dedicated to disciplinary teaching in Europe.

[5] The Treaty of Osimo deeply affected Trieste and its people. The treaty officially recognized the defacto border between Italy and Yugoslavia, ending the hope to recover at least a portion of the territory lost in WW2. These territories were important parts of Trieste's regional economy; their loss condemned the city to a smaller economic footprint. Second, Italy recognized Yugoslavia's nationalization of the properties of Italian citizens who had left the former Italian lands, many of which lived in Trieste and surrounding towns. A sizeable portion of Triestino voters lost hope to recover their properties and ancestral homes. Strong reactions were the obvious consequence.

contributing to the most pressing issue affecting the city—the reorganization of the border and borderland communities. Two articles were published that attempted to summarize the new orientations in the study of boundaries (Pagnini 1976; Valussi 1976). These two articles were foundational for future work by *Triestino* geographers by developing the legacy of the founder, Roletto, in separate directions. Pagnini's work is synthetic and mainly abstract. She begins by affirming "in reality, there is no defined space" and seeks a new interpretation by referring to theories of perception and structuralism. Valussi's essay on the other hand is broader and more thoughtful. Aware of the "defunctionalization now underway of many borders within Europe", he adopts a perspective that "focuses on borders as factors of the territorial organization". Taken together, their contributions fit both in the framework of classical human geography and of a renewed political geography. The analysis of borders and borderlands necessarily leads to regional research in which the systematic approach (the nature of the border and its fixing process) and the regional approach (the effect of the border on the frontier zone) are complementary. Hence the emergence of the borderland, which is defined as "the area where phenomena induced by the presence of the border take place," as the privileged object of geographical studies.

1976 was also a crucial year for the city when the Treaty of Osimo brought again geopolitics to the fore of public discourse in the city. My contribution to *Triestino* geography started there, combining the approach of my mentor, Bonetti, who was introducing the quantitative revolution in Trieste,—with the examples offered by the territory: in a nutshell, I was the last of a line of geographers that were continuing Roletto's search for a new balance between cities and regions, concerning the reality of Trieste. My work, the work of Valussi, Bonetti, Pagnini, and the others,

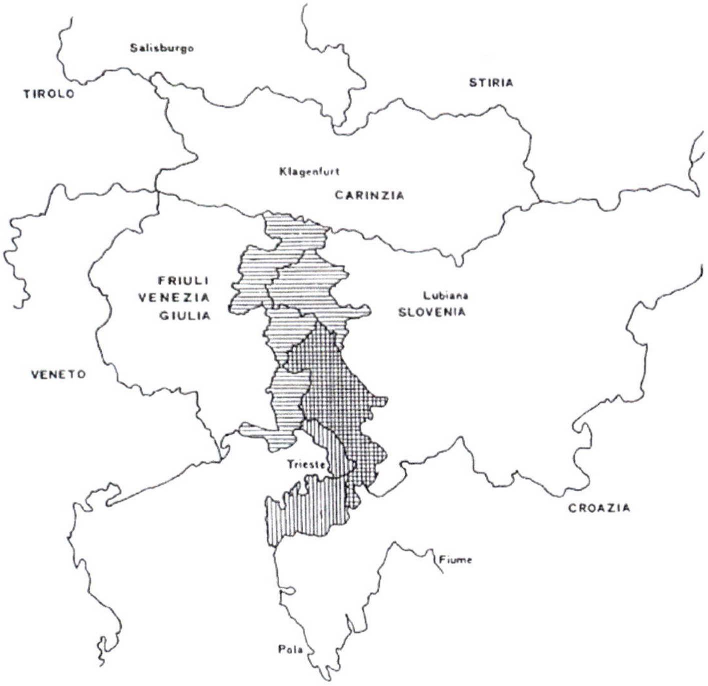

Fig. 4.2 Italian-Yugoslav cross-border areas with facilitated commercial traffic: clearing agreements for Gorizia/Nova Gorica and Trieste/Capodistria (Koper). *Source* Battisti (1984)

all the way back to Roletto was centered upon urban and regional geography,[6] as shown in Fig. 4.2, one of my earlier studies of the political economy of cross-border trade. However, the underlying themes were quintessentially political, because at the time the world of politics had the will and possibility of imposing territorial structures through the intersection of the 'grand' geopolitics of the Cold War and the micro-dimension of urban politics.

Although Trieste was a special place to study the intersections between regional and political geographies, it was not the only one. Among Italian geographers, the first, technocratic, studies of the intersection between politics and regional geography started with studies on territorial assets in the context of national planning (Compagna 1967; Muscara 1968; Corna Pellegrini 1974, Bagnasco 1977). Through these studies, we became more aware of the political element as a "motor" of the transformations of reality, an area of inquiry in international geography since 1967 (Gambi 1968; Lefebvre 1972; Harvey 1973), further developed in the 1980s by works on the "geography of power" (Raffestin 1981, 1983; Racine et al. 1978).

4.4 Falling Walls: *Triestino* Geography at the End of the Bipolar World

While the world was moving toward the end of the Soviet Union and the bipolar world, international human geography was more and more interested in the intersections between place, meaning, and power; in a nutshell, it was experiencing the 'cultural turn.' This is a direction that Valussi could not follow, faithful to the traditional conception of scientific research; far from partisan compromises he was accustomed to a geography strongly linked with the concrete features of the territory. Others attempted to accommodate the new trends alongside the more traditional legacies of our school of thought (Pagnini 1987). Over time, *Triestino* geography began to include the prevailing themes in international geography. For example, Zanetto,[7] influenced by the French geographer Breton (1978, 1988), pushed forward discussion of the nature of power through studies of language minorities (Zanetto 1983, 1987).

Besides the cultural turn, even before the fall of Socialist regimes in 1989, international political geography asked questions about changing territorialities, detecting the disruption of consolidated geopolitical structures (Smith 1984; Strassoldo 1981). These changes deeply affected the urban and regional geographies around Trieste: for example, north of Trieste, the town of Gorizia, split by the Iron Curtain (Fig. 4.3), could begin the reconstruction of its urban space (see Chap. 6). The process of European unification (Lizza 1991) contributed, together with the work of multinationals (Taylor and Thrift 1981), to shifting great powers' confrontations from land to the oceans (Vigarié 1990). At the micro-level, these analyses included a renewed attention to cross-border regions, i.e. functionally connected, but non-homogeneous regions. In Trieste, our position and tradition led us to pay particular attention to the latter issue: Valussi's last work, published posthumously, investigated the cross-border arrangements of the Alpine region (1991).

Valussi passed away right at the end of the confrontation between the USA and USSR that had shaped our city so profoundly.[8] The passing

[6] As for the scientific contribution of the masters of Triestino geography—Roletto, Bonetti, Cucagna, Valussi, Lago—see the obituaries in the bibliography Battisti (1988) and Battisti (2006, 2018, 2020)

[7] One of the best scholars of my generation, Gabriele Zanetto came from Venice to briefly hold the chair that had been Bonetti's. Unfortunately for Trieste, he almost immediately returned to his home university, abandoning a promising line of research of particular interest for our area.

[8] Pio Nodari, Valussi's student, passed part of Valussi's intellectual legacy and research program to his student, Francesca Krasna, who in the end inherited Valussi's professorship. In addition to continuing Valussi's research on migrations (Krasna 2009), Krasna's research focuses on the dissolution of Yugoslavia and the new order of the Karsna (1977, 2002).

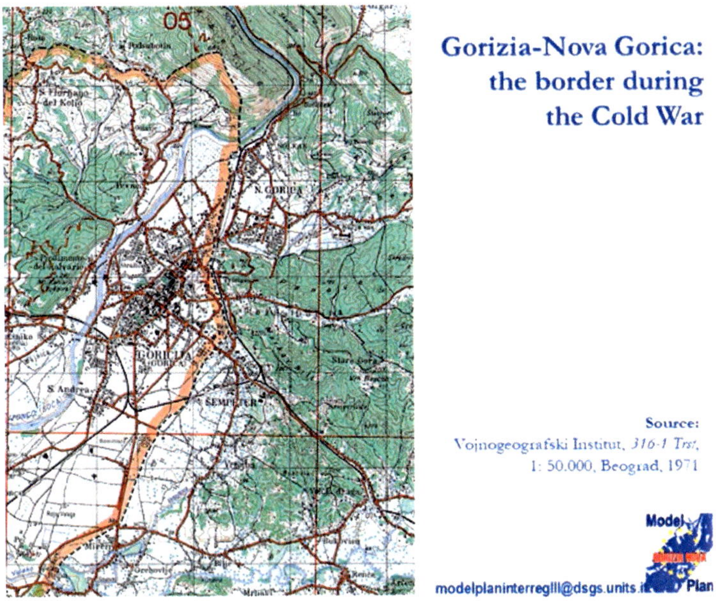

Fig. 4.3 Gorizia: a "cold war" border. *Source* ModelplaninterregIIIa

of Valussi occurred in a season marked by the growth of international relations, which included increased contacts with foreign scholars. The internationalization of Triestino geographers built on solid bases, in part thanks to my work. Already in 1983, in Oxford UK, I had the opportunity to participate in the founding meeting of the Working Group of the International Geographical Union (IGU) on the World Political Map. Unfortunately, in the 1990s I had to forego many international opportunities because I took an increased administrative and teaching load, as both program director of the major in Communication and as chair of the Department of Geographical and Historical Sciences. It was Pagnini's turn to lead the internationalization effort, mostly at the IGU Commission on Political geography.

Established in 1991, the Department of Geographical and Historical Sciences gathered most Triestino geographers. We had enough faculty members to establish, in 1995, the Ph.D. program in geohistory and geoeconomics of border regions, which I initiated. Numerous talented scholars emerged from the program. A few of them built careers publishing in geopolitics, including Prof. Francesca Krasna and the coauthor of this volume, Prof. Christian Sellar. A few years earlier, Prof. Pagnini initiated, almost alone a similar Ph.D. program in Political geography within the Department of Political Science, drawing on faculty members from various Italian universities, with an affiliated branch at the University of Naples.

Unfortunately, 'my' doctorate closed in 2007, anticipating the closure of the whole Department by a few years, because of one of many reforms of the Italian university system. Pagnini instead was able to resist until she moved to the Telematics University "Nicolò Cusano" in Rome, where she re-founded her Ph.D. program, which still continues under her guidance. Besides Pagnini, there was another scholar in Rome, Lizza, who also published in geopolitics (1996, 1999, 2001, 2004, 2009, 2021). Lizza was a student of Massi, further highlighting the importance of the *Triestino school* in this field of Italian geography.

In short, notwithstanding the sometimes unfavorable reforms of the university system, the closure of the Ph.D. program, and the transfer of many talented geographers outside Trieste, the *Triestino School* persists. The growing European integration culminating with the accession of Slovenia (2004) and Croatia (2013) to the European Union transformed the geopolitical role of Trieste from the strategically relevant and economic marginal frontier of the West to one of the many inner areas of the EU. Free from the civic duty to speak on behalf of a city under existential threats, *Triestino* geographers could expand their analyses, asking old questions about the most suitable organization of territories within newer theoretical approaches. Some, like myself, included new themes but remained faithful to the realist, organic approach that dates back to our founder, Roletto. Such a long legacy is the main reason why the following chapters will start with historical geographies of Trieste dating back more than a century (Chap. 5), to root questions about administrative divisions in a contemporary transborder community (Chap. 6) in the deep past. I will then scale up to questions about the geopolitics of integrated Europe and the new world order emerging in the twenty-first century by integrating the spatial structures of politics with those of the economy. In short, my work is the continuation of a century-long, epistemologically distinctive, geographical tradition that never failed to speak on behalf of a city oftentimes at the center of grand geopolitical games.

References

Adami V (1931) Storia documentatadeiconfini del Regno d'Italia [Documented history of the Kingdom of Italy], vol 4. Confine italojugoslavo, a cura dell'Uff. Storico del Ministero della Guerra, Roma, 1st. Poligr. dello Stato, pp XI–1058

Agnelli A (1971) La genesi dell'idea di Mitteleuropa [The origin of the idea of Mitteleuropa]. Milano

Antonsich M, Kolossov V, Pagnini MP (2001) On the centenary of Ratzel's Politische Geographie. Europe Between Political Geography and Geopolitics, Roma, Società Geografica Italiana

Ara A, Magris C (1982) Trieste un'identita difrontiera [Trieste, a frontier identity]. Torino, Einaudi

Bagnasco A (1977) Tre Italie. La problematica territoriale dello sviluppo urbano [Three Italies. The territorial issues of urban development]. Bologna, Il Mulino

Barbi P (1975) Un errore politico La cessione della "Zona B" (A political mistake the cession of "Zone B" (Speech to the Chambre of Deputies on 2.10 1975, brochure without date and place of printing)

Battisti G (1984) La collaborazione del Friuli-Venezia Giulia con la Jugoslavia e le prospettive di sviluppo, Excerpt from: Atti 8° Convegno Scientifico Internazionale Alpe-Adria (Graz, 7–8 ottobre 1982). Facoltà di Economia e Commercio, Università di Trieste, pp 107–223

Battisti G (1988) Alessandro Cucagna (1917–1987). Bollettinodella Societa GeograficaItaliana 11(5):471–473

Battisti G (2006) Eliseo Bonetti: ritratto di un geografo [Eliseo Bonetti potray of a geographer]. Rivista Geografica Italiana 63(2):367–369

Battisti G (2018) Luciano Lago (1937–2017). Bollettino della Societa' Geografica Italiana 14(1):231–237

Battisti G (2020) Valussi Giorgio, in Dizionario biografico degli italiani. http://www.treccani.it/Portale/ricerche/searchBiografie.html

Bettiza E (1966) Mito e realtà di Trieste [the myth and reality of Trieste]. Milano, All'insegna del pesce d'oro

Bevilacqua G (1984) La minoranza slovena a Trieste e il rapporto Italia—Yugoslavia [The Solvienian ethnic minority and the relations between Italy and Yugoslavia]. Cenni di diritto e di storia, Trieste, LINT

Bonetti E (1960) Gli sviluppi dell'insediamento nel bacino del Fella, con particolare riguardo all'area linguistica mista [Settlement developments in the Fella basin, with particular regard to the mixed linguistic area]. Pubbl. dell'Ist. di Scienze Politiche dell'Univ. di Trieste, Trieste, Smolars, p 128

Breton R (1978) Geografia delle lingue [The geography of language]. Venezia, Marsilio

Breton R (1988) Etnie. Paesaggio umano [Ethnic groups, human landscape]. Torino, Ulisse Edizioni

Cecovini M (1977) Discorso di un triestino agli italiani e altri scritti politici [Speech of a Triestine to the Italians and other political writings]. Trieste, LINT. Claval P (1978) Espace et pouvoir. Parigi, P.U.F.

Compagna F (1967) La politica della citta [The politics of city]. Bari, Laterza

Corna Pellegrini G (1974) Geografia e politica del territorio. Problemi e ricerche [Geography and politics of territory. Problems and research]. Milano, Vita e Pensiero

Cucagna A (1958) Considerazioni geografiche sul tracciato di alcuni limiti amministrativi delle Alpi Occidentali italiane [Geographical considerations on the layout of some administrative limits of the Italian Western Alps]. Rivista Geografica Italiana, Firenze, pp 144–150

Cumin G (1933) Appunti geografici sulla giunzione di frontiera della Venezia Giulia e sul confine italo-jugoslavo [Geographical notes on the Venezia Giulia border crossings and the Italian-Yugoslav border]. Porta Orientale, Trieste, Monciatti, Ill, pp 568–594

Cusin F (1937) Il confine orientale d 'Italia nella politica europea del XIV e XV secolo [Italy's eastern border in 14th and 15th Century European politics]. Milano, Giuffre

De Marchi L (1929) Fondamenti di geografia politica. Basi geografiche della formazione e dello sviluppo di stati e dei problemi politici attuali [Fundamentals of political geography. Geographical bases of the formation and development of states and current political problems], 2nd edn. Padova, Cedam (1938)

Gambi L (1968) Geografia e contestazione [Geography and contestation]. Faenza, F.lli Lega

Gerin G (1952) Considerazioni giuridiche sulle vicende del Territorio di Trieste [Juridical considerations on the events of the Territory of Trieste]. In: Annali Triestini, Univ. di Trieste, vol XXII, Sez I, Trieste

Gerin G (1953) Una interessante sentenza della Corte di Cassazione sullo status del Territorio di Trieste [An interesting ruling of the Court of Cassation on the status of the Territory of Trieste]. In: Rassegna giuliana di diritto e giurisprudenza, no 3–4, Trieste

Gerin G (1973) La giurisprudenza sul/a posizione delle cosiddette zone A e B del mai costituito Territorio Libero di Trieste. Edizioni della "Rassegna giuliana di diritto e giurisprudenza", Trieste

Harvey D (1973) Social justice and the city, London, Edward Arnold

Haushofer K (1925) PolitischeErdkunde und Geopolitik [Political geography and geopolitics]. Monaco, Oldenbourg

Haushofer K (1927) Grenzen in ihrergeographischen und politischenBedeutung [Borders in their geographical and political meaning]. Bertino K., Vowinckel Verlag

ISDEE (1989) Venti anni di attivita dell'ISDEE [Twenty years of ISDEE's activity]. Trieste, ISDEE

ISIG (1990) Venti anni dell 'Istituto di Sociologia lnternazionale di Gorizia 1969–1989 [Twenty years of the Institute of International Sociology of Gorizia]. Gorizia, Grafica Goriziana

Jaja G (1939) E' lo spirito che doma e piega la materia. Confidenze ai geografi. [It is the spirit that tames and bends matter. Confidences to geographers]. In: Bollettino della Societa Geografica ltaliana, XVII:124–142

Kasperson RE, Minghi BV (1969) The structure of political geography. Aldine Publishing Co., Chicago

Krasna F (1977) Nascita e Morte di uno Stato. Un'Interpretazione Geopolitica della ParabolaJugoslava [Birth and death of a state. A geopolitical interpretation of the trajectory fo Yugoslavia]. In: Quaderni del Centro Studi Economico-Politici 'Ezio Vanoni', no 1–3, Trieste, La Mongolfiera

Krasna F (2002) Ripensare i Balcani nel nuovo scenario geopolitico mondiale [Rethinking the Balkans in the new global geopolitical scenario]. Università degli Studi di Trieste—DSGS, Trieste

Krasna F (2009) Alla ricerca dell'identità perduta. Una panoramica degli studi geografici sull'immigrazione straniera in Italia [The search of a lost identity. A review of geographical studies of international immigration in Italy]. Bologna, Pàtron

Lago L (1966) Tarvisio: analisi di un centro di frontiera [Tarvisio: analysis of a frontier center]. In: Rivista Geografica ltaliana, Firenze, pp 488–493

Lefebvre H (1972) Le droit a la ville. Espace e politique [The right to the city. Space and politics]. Parigi, Anthropos

Lizza GF (1991) lntegrazione e regionalizzazionenella CEE [Integration and regionalization of Central and Eastern Europe]. F. Angeli, Milano

Lizza G (1996) Territorio e potere. Itinerari di geografia politica [Territory and power. Itineraries of political geography]. Torino, UTET

Lizza G (1999) Geopoliticadella nuova Europa [The geopolitics of new Europe]. UTET, Torino

Lizza G (2001) Geopolitica [geopolitics]. UTET, Torino

Lizza G (2004) Paneuropa [pan-Europe]. UTET, Torino

Lizza G (2009) Scenari geopolitici [geopolitical scenarios]. UTET, Torino

Lizza G (2021) Gli orizzonti della nuova geopolitica. Verso il 2050 [The horizons of the new geopolitics. Towards 2050]. UTET, Torino

Lokar A (1971) Il mito di Trieste [The myth of Trieste]. In: Most, Trieste, pp 77–79

Magris C (1980) Danubio [Danube]. Milano, Garzanti

Marsico P (1979) L 'Italia e la conferenza economica di Portorose, 24 ottobre–23 novembre 1921 [Italy and the eocnomic conference of Portorose, October 24–November 23 1921]. Coll. Fac. di Scienze Politiche, Univ. di Trieste, no 17, Milano, Giuffre

Maserati E (1963) L 'occupazione jugoslava di Trieste, maggio-giugno 1945 [The Yugoslav occupation of Trieste, May–June 1945]. Del Bianco, Udine

Massi E (1986) Geopolitica: dalla teoria originaria ai nuovi orientamenti [Geopolitics: from the original theory to new directions]. In: Bollettino della Società Geografica Italiana, 1–6, pp 3–45

Migliorini E (1932) Aspetti geografici efattori politici nella contesa fra Danzica e Gdynia, [Geographical aspects and political factors in the dispute between Gdansk and Gdynia]. In: Bollettino della Societia Geografica ltaliana, Roma, pp 3–49

Muscarà C (1968) Una regione per il programma [One region for the program]. Padova, Marsilio

Pagnini MP (1976) Sul concetto di confine: nuovi orientamenti metodologici [On the concept of borders: new methodologies. Raccolta Scritti 50° anniversario dell'Universita' di Trieste, pp 121–129

Pagnini MP (1987) La geografia politica [political geography. In: Aspetti e problemi della geografia, Milano, Marzorati, vol I, pp 409–442

Pagnini MP (ed) (2011) In: Proceedings of the 2nd conference on the adriatic forum, construction and deconstruction of nationalism and regionalism. The long journey to Europe, Montpellier, Academie Europeenne de Géopolitique

Pagnini MP, Sanguin A-L (2015) Storia e teoria della geografia politica. Una prospettiva internazionale [History and theory of political geography. An international perspective]. Roma, Edicusano

Pupo R (1979) La rifondazione della politica estera italiana: la questione giuliana 1944–46 [The refoundation of Italian foreign policy: the Julian question 1944–46]. Udine, Del Bianco

Quarantotti-Gambini PA (1951) Primavera a Trieste [Spring in Trieste]. Milano, Mondadori

Racine JB, Raffestin C, Ruffy V (1978) Territorialita e paradigma centro-periferia. La Svizzera e la Padania [Territoriality and the center-periphery paradigm: Switzerland and Padania]. Milano, UNICOPLI

Raffestin C (1981) Per una geografia del potere [For a geography of power]. Milano, Unicopli

Raffestin C (1983) Geografia politica: teorie per un progetto sociale [Political geography: theories for a social project]. Milano, Unicopli

Roletto G (1915) Termini geografici dialettali delle Valli Valdesi [Geographical dialect terms of Valli Valdesi]. In: Rivista Geografica Italiana, Firenze, pp 191–199 e 285–293

Roletto G (1932) Sulle nuove unita amministrative delle Alpi Occidentali. In: Rivista di Geografia, Firenze, pp 176–180

Roletto G, Massi E (1931) Lineamenti di Geografia politica. Introduzione, Parle I: i confini [Introduction to political geogrpahy. Introduction, part 1: the borders]. Trieste, Pubblicazioni dell'Istituto di Geografia della R. Universita

Ronchey A (1973) Atlanteideologico [The ideological atlas]. Milano, Garzanti

Sambri C (1970) Una frontiera aperta. Indagine sui valichi italo-jugoslavi [An open frontier. A study of Italian-Yugoslav border crossings]. Bologna, Forni (Pubbl. ISIG, "Serie ricerche", no 1)

Schiffrer C (1946) La Venezia Giulia. Saggio di una carta dei limiti nazionali italo-jugoslavi [Venezia Giulia. Essay of a map of the Italian-Yugoslavian national limits]. Roma, Colombo

Schiffrer C (1978) Le origini dell'Irredentismo Triestino (1813–1860) [The origins of Trieste's Irredentism]. Trieste, Del Bianco, Roma, Colombo

Smith A (1984) Il revival etnico [The ethnic revival]. Il Mulino, Bologna

Strassoldo R (1981) Civiltà Mitteleuropea [Central european civilization], Geschichte undSoziologieeinerBewegung, AAVV, Mitteleuropa, Hofgeismar, Evangelische Akademie

Taylor PJ, Thrift N (1981) The geography of multinationals. Croom Helm, London

Tomizza F (1976) La miglior vita [The better life]. Milano, Rizzoli

Turco A (ed) (1984) Regione e regionalizzazione [Region and regionalization]. Milano, F. Angeli

Udina M (1969) Scritti sulla questione di Trieste sorta in seguito al secondo conflitto mondiale ed i principali atti internazionali ed interni ad essa relativi [Essays on the question of Trieste that arose after the second world conflict and the main international and internal acts related to it]. Milano, Giuffre, p 466

Valdevit G (1984) La questione di Trieste 1941–1945. Politica internazionale e contesto locale [The Trieste question 1941–1945. International politics and local context], Milano, F. Angeli

Valussi G (1957) L'influenza delle variazioni di confine sul turismo nella Venezia Giulia [influences of border shifts on tourism in Venezia Giulia]. In: Bollettino della Societa' Geografica Italiana, Roma, pp 412–425

Valussi G (1962) Aspetti geografici di una vecchia lite fra due comunita prealpine (Erto e Casso) [Geographical features of an old dispute between two pre-alpine communities, Erto and Casso]. In Ce fastu?, Udine, pp 103–116

Valussi G (1965) L'opera scientifica di Giorgio Roletto [Giorgio Roletto's scientific work]. In: Bollettino della SocietA Geografica Italiana, Roma, pp 313–326

Valussi G (1971) Il riassetto amministrativo ed economico dei territorio della Venezia Giulia annessi alla Repubblica Popolare Slovena [The administrative and economic reorganization of the territories of Venezia Giulia annexed to the People's Republic of Slovenia] in Miscellanea I, Udine, Arti GraficheFriulane, pp 243–265 (Pubbl. Fac. di Lingue dell'Univ. di Trieste, 1)

Valussi G (1972) Il confine nordorientale d'Italia [Italy's northeastern border]. Trieste, Lint

Valussi G (1976) Nuovi orientamenti nella geografia dei confini politici [new trends in the geography of political boundaries]. In: Rivista Geografica Italiana, Firenze, pp 41–52

Valussi G (1991) Le intese transfrontaliere della Regione Alpina: la Comunita di Lavoro Alpe-Adria [Cross-border Arrangements in the Alpine Region: the Alps-Adriatic Working Community]. In: Bollettino della Societa Geografica Italiana, Roma, pp 443–455

Vigarié A (1990) Geostrategie des oceans [The geostrategy of oceans]. Caen, Paradigme

Zanetto G (1983) Il Quebec: geografia di una lingua [Quebec, the geography of a language]. Scuola Superiore di Lingue Moderne, Trieste

Zanetto G (1987) Lingue e geografia: l'etnoregionalismo [languages and geography: ethnoregionalism]. In: Corna Pellegrini G (a cura di) Aspetti e problemi de/la geografia, vol I, Settimo Milanese, Marzorati, pp 445–463

Part III

Historical Geography as Method: Producing Geopolitics from the Julian Region

Check for updates

5 Urbanization Processes in a Transnational Area. An Application of the Rank-Size Rule to the Austrian Littoral

Abstract

This chapter uses the rank-size rule (RSR), one of the mathematical tools developed during the quantitative revolution, to tackle questions of territorialization and reterritorialization that are highly relevant in contemporary human geography. It asks the question of the persistence of territory, understood as space framed by political organization, after the framing agent ceases to exist. Specifically, it looks at the Austrian Littoral, a subnational administrative unit of the Austrian Empire that existed from 1849 to Austria's dissolution in 1918, which is now divided among the countries of Italy, Croatia, and Slovenia. By comparing the 1910 Austrian census with 2001 data, the chapter concludes that the Littoral never had the integrated territorial structure predicted by the RSR, and that even older territorialities shaped the region. It also concludes that the infrastructural projects initiated at the end of the twentieth century are just now beginning to change the nature of the region's urban hierarchy.

Keywords

Rank Size Rule (RSR) · Austrian Littoral · Territorialization · Census data · Long term analyses

5.1 Introduction—Some Issues with Territories

Consciously or unconsciously, the work of the geographer concerns the definition of regional spatial entities, which means, more often than not, their redefinition (Raffestin 1984). We geographers are aware that all territorial structures are subject to constant change, which affect their character as well as their extension. Therefore, we need a methodology that allows us to monitor change over time, to update and redraw the localization of any given entity. To be effective, such a task requires studying territories over long periods. In this essay, I adopt economists' definition of long time frames (Marshall 1890), i.e. the period firms need to modify qualitatively as well as quantitatively all factors of production they use.

Because territories are complex spatial entities and not simple firms, it is necessary to consider the entire stock of firms in a given space and thus extend the length of time frames even further than what economists would suggest. For example, certain enterprises (such as those involved in exploiting oil fields) have an economic cycle of over thirty years. Thus, the geographical 'long time frame' involves the time necessary for a radical transformation of the entire economic context in a given area.

C. Sellar and G. Battisti, *Geopolitical Perspectives from the Italian Border*, Historical Geography and Geosciences, https://doi.org/10.1007/978-3-031-26044-5_5

The adoption of long-term views allows testing the resilience of territorial structures, which in turn allows judging the efficacy of regional subdivision. This is especially relevant for political and administrative units. In terms perhaps more familiar to Anglo-American colleagues, this chapter argues that long-term analyses, in the time frame of decades or centuries, provide a better understanding of deterritorialization and reterritorialization. Long-term analyses help understanding territory in terms of resilience, because a single, albeit traumatic, event is usually not sufficient to break down a proven spatial subdivision (Capineri et al. 2014).

The concept of resilience of territorial structures allows for both micro- and macro-level analyses. The classical method of geography consists of place-specific micro-level analyses, leading inductively to generalizable conclusions. Vice-versa, deductive analyses at the systemic level lead to syntheses, highlighting the essential elements of reality.[1] However, in both cases the primary goal of the analysis is to identify the boundaries of the space under consideration. These boundaries are often coincident with, and legitimized by, administrative units in the political system (cities, districts, provinces, states, and the like). Legitimacy is oftentimes reinforced by the government-led collection of statistical data: governments collect data within the administrative units they devise and later use the data to normalize the existence of the units, in a circular, self-referencing process.

To better understand resiliency, this chapter analyzes a territorial entity with a long history, which was subjected to a series of shocks that caused its dissolution as an administrative unit. The key research question is: to what extent the disbandment of the administrative structures corresponded to a systemic change in the underlying socioeconomic and urban structures? The following case study shows that administrative dissolution by placing a new international border in the middle of an earlier province did not cause radical changes in regional economic structures and spatial patterns of development, although it led to profound ethnocultural transformations. Such findings led to a further question; i.e. what is the role of changing national borders on the transformation of this specific territory?

5.2 The Area Under Investigation

To better understand the long-term territorial dynamics mentioned above, this chapter focuses on the Austrian Littoral (Fig. 5.1), officially established as a province of the Austrian Empire in 1849; after WW1 it will be annexed by Italy, and, with some small changes, it will maintain its status as a discrete administrative unit with the new name *Regione Venezia Giulia*[2] (aka Julian Region). Following WW2, the peace treaty of 1947 and the London memorandum of 1954 split the region: Italy kept a small portion, including the capital city Trieste, which was incorporated in the new administrative unit *Friuli Venezia Giulia*, while Yugoslavia annexed the overwhelming majority of its land. Within Yugoslavia, the newly acquired land was further split between the republics of Croatia and Slovenia, which still keep it after they became independent nations following the breakup of Yugoslavia in 1991. Thus, with different names, the same subnational administrative unit persisted for a century while shifting to different countries, to the breakup into three separate pieces belonging to three different countries during the eventful twentieth century.

[1] The opposition between inductive and deductive approaches is a false dichotomy. In a letter to N.L. Keynes (John Manyard Keynes' father), A. Marshall stated: "You talk of the inductive & the deductive method, whereas I contend that each involves the other, and that historians are always deducting, & that even the most deductive writers are always implicitly at least basing themselves on observed facts." (Groenewegen 2013).

[2] The name *Venezia Giulia* predated the renaming of the Austrian Littoral: it was first used in 1863 by the glottologist G. I. Ascoli to classify the Italian-speaking areas of the Hapsburg Empire: *Venezia Giulia* identified the easternmost areas, surrounding the eastern portion of the Alps, *Venezia Euganea* the plains surrounding Venice, and *Venezia Tridentina* the alpine areas north of Lake Garda. His classification is still today the basis of three of Italy's provincial and regional administrations (Salimbeni 1991a).

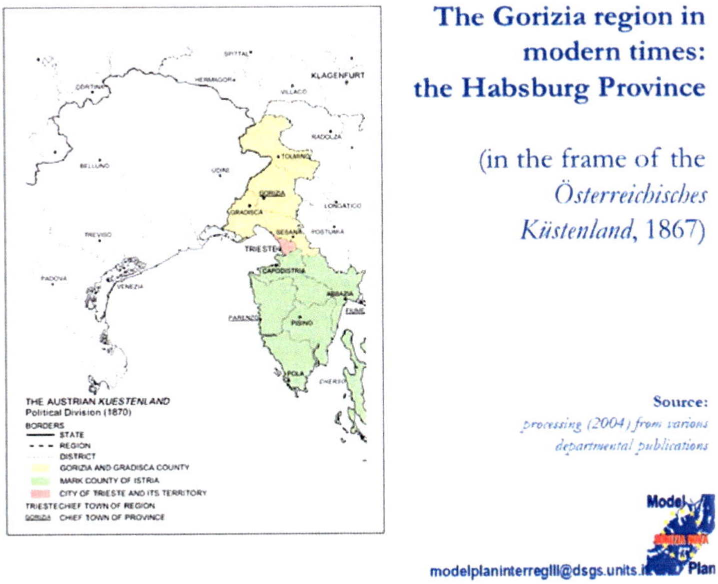

Fig. 5.1 The Austrian Littoral province. *Source* ModelplaninterregIIIa

Some scholars pointed out that the Austrian Littoral, as a discrete administrative unit, existed even before 1849 (Cervani 1980; Salimbeni 1991b). It represented the bulk of the Kingdom of Illyria, a crown land of the Hapsburg Empire, established in 1816 after Austria annexed the Illyrian Provinces, the French territory Napoleon has established in 1809 as a buffer between France, Austria, and the Ottoman Empire (Pivec-Stelé 1930). The main geographical difference between the Kingdom of Illyria and the later Austrian Littoral is that the latter lost the areas surrounding the cities of Carlstadt and Fiume (today Rijeka). These were kept in Hapsburg's hands but were attributed to Hungary through the *Ausgleich* of 1867, when they restructured their empire to constitute the Austro-Hungarian dual monarchy. Fiume briefly returned to the Littoral (aka *Venezia Giulia*) in 1924 when Italy annexed it. However, outside the old Hapsburg space, it lost the role of Hungary's largest port, thus losing its traditional economic function.

In a nutshell, as a subnational unit of the Hapsburg Empire, the Austrian Littoral existed for nearly 70 years. Adding the Italian period, it reached the century. Considering the earlier Napoleonic province, it lasted 130 years. More than 75 years have passed since its partition after WW2, two centuries have passed since its origin; it is enough time to perform the long-term analysis discussed in the introduction, aimed at identifying the elements of unity persisting in its former spaces. The goal of the analysis is not to support the nationalist agenda of any of the five countries mentioned above, but to better understand the dynamics of an area whose territorial subdivisions have been exceedingly fleeting since WW1 (Battisti 1979). The multiplicity of cross-border cooperation programs that started in the 1960s and intensified after Slovenia and Croatia joined the EU is a further demonstration of the fragmentation of a substantially unified space.

5.3 Methodology

There is a rich academic literature—for the most part in Italian—that discusses the Littoral from a variety of disciplinary standpoints: literary studies, history,[3] linguistics, sociology, law, economics, and ecology. Geographers have contributed as

[3] The journal *Quaderni Giuliani di Storia* [Julian History Notebooks] publishes historical research on the region.

well, focusing on some highly sensitive, policy relevant topics. These included the 'Trieste question,' i.e. the territorial dispute between Italy and Yugoslavia, which was Italy's most important geopolitical flashpoint for two decades, between the end of WW2 in 1945 and the peace treaty the two countries signed in 1976 (Valussi 1972; Battisti 1979). Although Trieste was the most important prize in the dispute, because it was the largest city in the area as well as headquarter of several industrial conglomerates that had grown around its port, the two countries were competing to secure the entire former Littoral.

Practically speaking, the two governments were arguing over the border, with different interest groups drawing a variety of proposals for the new borderline, each claiming they were using 'scientific' criteria.[4] The problem of the line—both as a barrier and as the identifier of cross-border points—focused scholars' attention away from broader regional issues. Over time, regional studies of the region became less and less up to date, largely because of the disappearance of the Littoral/Venezia Giulia as a discrete administrative unit. Thus, there is an open space for inquiry scholars should pay attention to, possibly against the grain of international geography, which increasingly considers regional monographs as obsolete.

Lack of interest in regional monographs does not mean lesser production of knowledge: today geographers have better methods to describe and interpret territories than they had at the time of the territorial disputes between Italy and Yugoslavia, when French regional geography and landscape studies still were the dominant theoretical paradigms. In Italy, the study of urban networks is a widely accepted tool for comprehensive regional studies; however, the split of the former Littoral's urban areas among different countries, compounded by fifty years of opposing ideologies and economic systems—capitalist Italy and socialist Yugoslavia—further complicated this research.[5]

To avoid some of the pitfalls of data collections across different systems and long time frames, this work uses the rank-size rule (RSR) to identify meaningful spatial and temporal intervals, which later research will use to analyze in-depth the structural aspects of each local system. RSR was initially developed as a tool to analyze the spatial concentration of human populations. Therefore, it does not discriminate the nature of settlements (i.e. urban vs. rural) or their functions. It is worth remembering that Christaller (1933) tied the notion of urban with the outward projection of functions which he called 'central' (e.g., hospitals often serve populations beyond the cities they are located in), while any empirical research must contend with a variety of typologies of settlements—*agrovilles*, company towns, mining towns, pioneer towns, tourist cities, dormitory towns—that taken together defy simple categorization. Moreover, the functions of settlements change over time—even more strikingly in the long periods considered in this paper. For example, in the seventeenth and eighteenth centuries, early economists considered the provision of items of conspicuous consumption to the landed gentry as the primary function of cities. Christaller—an agronomist by training—would have rejected such categorization, but in the twenty-first-century scholars are rediscovering similar roles for globalizing cities (Glaser et al. 2001). Such shifts in functions did occur in the area investigated in this research, especially in the inner Istrian Peninsula and nearby Karstic reliefs around the mid-nineteenth century.

One of the benefits of the RSR is that it allows monitoring, albeit roughly, structural changes in systems of settlements over time (Cori 1976). It is an appropriate tool to answer this chapter's research question because it provides quantitative

[4] The Italian point of view built on the ethnographic map developed by Schiffrer (Venezia Giulia 1946) 3 was republished together with the accompanying report in the anthology curated by Verani (1990).

[5] Earlier in my career, I had to abandon a research project on cross-border central places, because I could not harmonize Italian and Yugoslav statistical data. After all, the collecting agencies used spatial and functional criteria of the capitalist versus socialist system—too different from each other to allow a meaningful comparison.

evidence of systemic destructuring that followed territorial partition; moreover, it allows a system-wide investigation that does not depend on micro-level analysis of specific indicators in each locale. Jefferson's seminal work on the RSR has similar concerns (1939): his intent was not to identify new urban typologies such as the primate city; instead, he shared with Christaller a concern for the consequences of shifts in borders after WW1 and other conflicts.

RSR is not free of limitations. Scholars have criticized it for decades, scrutinizing both the rigor of its mathematical model and its explanatory power, questioning especially whether a theory developed in the 1930s and 1940s is adequate to twenty-first-century realities (Fonseca 1988; Tong Soo 2004). I recognize the validity of their arguments, especially for newer types of settlements, such as the low-density urbanism that started in American cities and is now being adopted in a variety of contexts. However, the RSR is still adequate to study traditional settlement patterns, such as those found in the Littoral in the nineteenth and twentieth centuries: exhibiting less suburbanization than the average European region, cities, and villages in the former Littoral are similar to those the Rule was originally intended for.[6]

Once established the validity of the RSR for the area under investigation, the final methodological issue was how to choose the appropriate statistical indicator for the settlements. In such a long-term analysis, the emergence of conurbations and other modifications in the settlements has changed their structure. To avoid—or at least minimize—comparisons between non-homogeneous spatial structures—this chapter does not look at the total population in each township, but only at the population of downtowns, i.e. the centrally located neighborhoods. In so doing, it is possible to compare the demography of spatial units that have undergone the least changes over time. Such methodological choice leads to underestimate the total population, especially in the larger cities. The loss in population may be statistically significant, especially if other settlements expanded their downtowns at the expense of the larger city, thus altering the relative sizes of populations in a given area. However, such process of population redistribution away from the larger city—the so-called counterurbanization—has happened only in Trieste, and not in the rest of the Littoral.

5.4 Regional Dynamics

From the standpoint of international politics, the former Littoral seems to have experienced in the last century a full cycle of transformations, culminating in 2013 with Croatia joining the European Union. For the Littoral, EU membership of Croatia resembles a return to the past. Up to the beginning of WW1 in 1914, the Triple Alliance between Italy, Germany, and Austria-Hungary allowed citizens of the three countries freedom of movement and residence in each other's territory, which was especially important in the Littoral, at the time border region between Italy and Austria-Hungary. The growing territorial fragmentation after WW1 created increasing obstacles to the movement of peoples and goods until the enlargement of the European Union re-established conditions similar to the pre-WW1 period. In a nutshell, in almost exactly one century the former Littoral went from a situation of open international borders, to close, to open again. The restoration of freedom of movement begets the question of the economic, urban, and social validity of the Littoral as an integrated administrative unit, both at the time of its dissolution and today.

On the eve of WW1, the Littoral's settlement structure was heavily polarized on Trieste, which, according to the 1910 census, had a population four times higher than the second-largest city, Pula (Pola in Italian). At a first sight, such distribution suggests the presence of a primate city, although in this specific case the unbalance in the distribution of the population was not caused by direct statistical effects of geopolitical changes, but by the indirect effect of Austria losing control of Venice, and thus

[6] The persistence of such conditions in the twenty-first century is less certain, especially when cross-border settlement processes are considered (Battisti 2014).

investing in Trieste to develop a unique port city (Battisti 1993).[7] Surprisingly, a century later the relative size of the two largest cities is still roughly the same, even though the entire context (both within the former Littoral and outside) has changed. Across the new international boundaries, settlements boundaries in the former Littoral were redrawn multiple times, following a trend of administrative mergers of smaller units, initiated by the Italian government (Schiffrer 1953), continued in the Yugoslav period (Valussi 1971), and recently reversed after the independence of Croatia and Slovenia.

As mentioned earlier, administrative restructuring occurred in a context of repeated geopolitical shifts with dramatic—and traumatic—impacts on the region. These were: (1) the dissolution of the Austrian Empire; (2) reorganization of the township and provincial administrations, as well as economic development initiatives in the Italian period; (3) split of the region between Italy and Yugoslavia after WW2, roughly along ethnic lines; (4) dual socioeconomic development policies (capitalist in Italy, state socialist in Yugoslavia) during the Cold War; (5) violent breakup of Yugoslavia and emergence of a new international border between Croatia and Slovenia to cut across the Istrian Peninsula; (6) in the current phase, deregulation and territorial re-organization of townships in a globalizing context.

A complete analysis of the effects of each of these events would involve an analysis of all census data in the post-WW2 period, consisting of census years 1948, 1951/53, 1961, 1971, 1981, 1991, 2001/2, and 2011/12 for the three countries, vastly exceeding the preliminary analysis this paper aims to accomplish. Instead, to focus on the cumulative effects of the events mentioned above, the author initially chose to analyze the earliest and the latest data points available, i.e. the Austrian census of 1910 and the censuses of Italy, Slovenia, and Croatia of 2011/12. However, the long time lag between the two data points raised the issue of data compatibility: in a century, methods of data collection and aggregation have evolved, challenging the comparison between the two data sets. Furthermore, the three different governments involved in the 2011/12 census used different methodologies: the Croatian and Italian censuses opted for a direct, capillary data collection directly from households; the Slovenian census chose an administrative collection, compiling demographic data obtained from local governments. The Croatian Statistical Office changed methodology between the 2001 and 2011 censuses, adopting a new definition of resident population, which is different from Italy's and Slovenia's. Finally, at the time of the data analysis, Italy had yet to release its data. Thus, we chose to compare the 1910 census with the 2001 censuses, which were conducted with similar criteria in all three countries. Albeit methodologically imperfect, such comparison allows showcasing the cumulative effects of five of the six geopolitical shifts discussed above, omitting only the too recent phase of deregulation and territorial reorganization.

5.5 Data Analysis

Table 5.1 shows the results of the analysis, focusing on the 20 largest settlements. Because of the complex history of the region, the official names of each changed over time to reflect ethnonational concerns. To ensure ease of identification, and because the authors are writing from an Italian perspective, Italian names are used. The three columns on the left represent population distribution in 1910, the three on the right

[7] Jefferson's original model assumed that, in an integrated geoeconomic system, settlements followed a linear distribution, with the largest city having roughly twice as many inhabitants as the second, etc. (1939). Primate cities are much larger than any other settlement. Jefferson explained their existence in geopolitical terms: either they are the outcome of large imperial structures (i.e. London was the primate city of England because it was the center of the much wider settlement network of the British Empire) or the result of territorial losses (Vienna after WW1 was a primate city because Austria had lost the other large cities with the demise of its empire). The RSR supposedly works also for subnational territorial units, but Trieste grew beyond its expected size because of the economic growth caused by the Austrian state-led development of its port.

Table 5.1 Ranking urban centers in the former Austrian Littoral. Population in the main settlements in 1910 (three columns on the left) and 2001/2 (three columns on the right). The middle column represents the succession expected by the theoretical model

Hierarchy	Population	Succession	Theory	Hierarchy	Population	Succession
1. Trieste	220.545	100,0	100,0	Trieste	196.512	100,0
2. Pola	58.562	26,5	50.0	Pola	58.594	29,8
3. Gorizia	30.995	14	33,3	Gorizia	35.105	18,2
4. Rovigno	12.323	5,6	25.0	Monfalcone	25.877	13,2
5. Pirano	11.457	5,2	20,0	Capodistria	23.726	12,1
6. Capodistria	8.993	4,1	16,7	Nova Gorica	13.491	6,9
7. Monfalcone	7.136	3,2	14,3	Rovigno	13.467	6,8
8. Dignano	6.087	2,7	12.5	Cervignano	11.206	5,7
9. Lussinpiccolo	5.530	2,5	11,1	Ronchi d. L	11.075	5,6
10. Muggia	5.437	2,5	10,0	Muggia	10.995	5,6
11. Gimino	5.169	2,3	9,1	Parenzo	10.448	5,3
12. Grado	4.721	2,1	8,3	Isola	10.381	5,3
13. Pisino	4.425	2	7,7	Pirano	9.935	5,0
14. Cherso	4.112	1,9	7,1	Albona	7.904	4,0
15. Cormons	4.166	1,9	6,7	Abbazia	7.850	4,0
16. Abbazia	3.828	1,7	6,2	Umago	7.769	3,9
17. Portole	3.371	1,5	5,9	Villa Opicina	7.570	3,8
18. Umago	3.219	1,5	5,5	Grado	7.014	3,6
19. Buje	3.170	1,4	5,3	Gradisca	6.391	3,2
20. Cervignano	3.078	1,4	5,0	Ajdovscina	6.373	3,2

Sources Author's elaboration of *Spezialortsrepertorium der österreichischen Ländern* (1918)
ISTAT (2001)
Italian census (2001), Slovenian census (2002) and Croatian census (2001)

2001, and the central columns represent the distribution expected by the RSR, expressed as percentages of the population of the largest city. Thus, the largest city is indicated as 100, the second largest as 50 because RSR expects the second city to have 50% of the population of the first. In short, the table shows the RSR does not reflect the population distribution in the region in both 2001 and 1910. Table 5.2 shows the variation between the rankings expected by the theory and the actual data (section A) and the shifts in the position of the various settlements between the two census years under consideration. Seemingly paradoxically, over time the population distribution became more similar to RSR's predictions, with a 3.4% *less* discrepancy between the theoretical model and the actual data in 2001 than in 1910. Only 5 out of 20 settlements maintained the same ranking; 9 left the ranking; one new center joined the group (Nova Gorica—see next chapter); 3 of the remaining settlements lost positions (Rovigno falls 3 positions back, Grado 6, and Pirano 8), and 4 gained (1 position for Capodistria and Umago, 3 for Monfalcone, 13 Cervignano).

Given the traditional RSR does not reflect the data, then it is necessary to investigate other possible explanatory models. As stated earlier, the primate city model does not work, even though Trieste is by far the largest settlement in

Table 5.2 Variations in the urban hierarchy in the former Austrian Littoral: 1910–2001/2

Hierarchy																			
1	2	3	4	5	6	7	8	9	10	11	12	13	14	15	16	17	18	19	20
A																			
Differences with respect to the distribution expected by the theory																			
In 1910																			
0	23.5	19.3	19.4	14.8	12.6	11.1	9.8	8.6	7.5	6.8	6.2	5.7	5.2	4.8	4.5	4.4	4	3.9	3.6
2001/2																			
0	24.8	16.5	11.8	7.9	9.8	7.5	6.8	5.5	4.4	3.8	3	2.7	3.1	2.7	2.3	2.1	1.9	2.1	1.8
Differences (±) 1910–2001/2 according to ranking																			
0	+1.3	−2.8	−7.6	−6.9	−2.4	−3.6	−3	−3.1	−3.1	−3	−3.2	−3	−2.1	−2.1	−2.2	−2.3	−2.1	−1.8	−1.8
B																			
Movements within the hierarchy																			
Number of retreats (−) / advancements (+) with respect to the distribution of 1910																			
0	0	0	−3	−8	+1	+3	**−13**	**−12**	0	**−10**	−6	**−8**	**−7**	**−6**	+1	**−4**	+2	**−2**	+12
With respect to the new entries in 2001/2																			
					+16			+13		+11	+10		+82			+5		+3	+2

Source Author's elaboration of the data reported in Table 5.1

Legend (terms in bold) Settlements out of the ranking in 2001/2. For them and the new entries the figures show the least theoretical changes. The figures in rows A represent the percentage difference between the theoretical model and actual data. For example, according to the model, the second largest city (Pula) is expected to be 50% of the population size of the first (Trieste). In reality, in 1910 it was 26.5%. Thus, row A indicates a discrepancy of 23.5 in column 2, for 1910, meaning that the population size of the second city (Pula) as a fraction of the first (Tireste) was 23.5% smaller than predicted by the model (50%-26.5% = 23.5%).

both censuses.[8] In 1910, instead of a ranking of settlements, data show an eschewed distribution with few cities concentrating most of the population (and wealth) and a relatively uniform distribution of the remaining towns. In 2001, ranking becomes more visible, but we are far from the model's expectations. The colonial model does not work either: typically, overseas colonies had a ranking of settlements reflecting the lower positions of the RSR, but lacked the largest cities, because these were located in the metropole.[9] Given the inadequacy of these widely used models, the notion of 'non-regionalized space' may work here. Turco developed such a concept to describe a distribution of settlements in a "broken path toward the formation of a regional system" or a "regress in the process of regional formation" (1984). The latter definition reflects a notion or region (with or without the political connotation of a territory) as a 'reality in becoming,' i.e. an oriented process that evolves as a result of the collective will of the community—a so-called geographic program (Muscarà 1968). In 1910 such a program had yet to show results, and in 2001 was replaced by a variety of more or less conflicting strategies.

The hypothesis of a 'geographic program' in which long-term efforts of community building are visible through the RSR was supported by a study of Italy's urban structures conducted in the 1970s by the Italian Committee of Geographers (Cori et al. 1976). The study showed that in the first century since the Italian reunification, during which the country developed a national economy and infrastructure, the country's urban network progressively aligned with the RSR. Attempts to scale down similar analyses to the subnational level showed the RSR reflected the historical legacies of pre-unitary states, with RSR-like settlement patterns reflecting the political map before reunification. Others, such as Trieste's current region, *Friuli Venezia Giulia,* established by merging provinces left to Italy after the territorial losses of WW2 with parts of the adjacent *Veneto*, are much less so, supporting this paper's use of the RSR to estimate regional socioeconomic integration in the historical constituency.[10]

The lack of a systemic structure of settlements (at least in the typologies evidenced by the RSR) in 1910, roughly a century after the French first established a distinct administrative unit in the space of the Littoral, and the persistence of such situation a century later, together suggest a specific case of 'resilience.' Usually, the notion of resilience is applied to relatively well-defined structures (such as institutions, networks of infrastructure, and organizations), while our data show a persistent lack of structure. From a functional standpoint, which should be the focus of political and economic geography, a distinct regional space in the area of the former Littoral seemingly never existed. Such finding is politically relevant because both the Italian-speaking and Slovenian-speaking communities have criticized the administrative subdivisions of Italy's eastern border region after WW1. Indeed, this analysis suggests that not only the Italian administrative restructuring of the former Littoral should be considered as an 'invented region' (Michieli and Zelco 2008), but the practice of drawing arbitrary administrative boundaries in the area goes deeper in the past.

The situation of the former Littoral is not exceptional: historically, a great variety of subnational administrative units were 'invented' by

[8] The border shift in 1945, which cut Trieste from its countryside should have led to a decrease in population and economic activities. Instead, the proximity of the socialist and capitalist economic system had an unexpected effect on the demographic side. First there was the exodus of the Italians from the lost territories, many of them choosing to live near the border. Second, a temporary economic boom affected the city in the 1960–80 period, due to international trade in households' goods. Simply put, families from Yugoslavia and even other countries in the Socialist bloc came to Trieste to buy consumer goods, primarily clothing and appliances. This injection of prosperity (along with state support to industrial and port activities) masked the decline of Trieste (Battisti 1979). This will become evident after the settlement of the border quarrel.

[9] The lack of a colonial-like urban network challenges historical analyses claiming that the rela*tive underdevelopment of the region was the outcome of occupation by external powers, namely* Venice in the middle ages and the Hapsburg in the modern period.

[10] Lando (2011) discussed the history of Italy's regions.

policymakers and intellectuals (Di Napoli 2014; Mogavero 2016; Rossi 2020). In Italy, geographers have frequently criticized the structure of the country's most important administrative subdivisions, the *regioni* (Muscarà 1968; Galluccio 1998; Lando 2011). However, these critiques should emphasize more the relatively recent history of the Italian State (160 years at the moment of writing) and the even more recent history of the *regioni* as decentralized units of government (60 years). The lack of a distinctive functional structure and also of identity is not only a characteristic of the *Venezia Giulia,* (the Italian name of the former Littoral) but also of nearly all other *regioni*, whose boundaries were drawn in the early twentieth century, building on oftentimes unconvincing historical and cultural legacies.[11] In the former Littoral, the tendency toward functional fragmentation is not only compounded by its current division among three countries, but also by its lack of any significant autonomy of government, because both the Austrian Empire and interwar Italy were highly centralized states. Thus the Littoral lacked any 'regionalizing effect,' i.e. the cumulative consequences of decisions in the areas of infrastructures and locations of economic activities, taken by a local government that makes plans on behalf of its territory, and it is only peripherally concerned with external interests.

The results of this preliminary research suggest that future investigations of the former Littoral should focus on different measures of territorial integration, such as the link between physical geography and settlement patterns, and the complex interactions between coastal areas and the interior. The author is committed to keeping working on the many geographies of this region.

5.6 Conclusions: The Many Reasons of Places

As stated earlier, even in 1910 the Littoral did not constitute a network of settlements within the parameters of the RSR. Such a conclusion will disappoint Trieste's many praisers of past times, but this is the objective finding of this study. Such finding also confirms that the formal establishment of an administrative unit is not sufficient to create an integrated socioeconomic space. In so doing, it supports the arguments of many competing nationalist ideologies of the nineteenth and twentieth centuries that, albeit opposed to one another, shared the complaint against the status quo of the Littoral. Moving beyond ideological struggles, a scholar's job is to investigate the underlying reasons for the apparent lack of integration.

Starting with the situation in 1910, the first explanation may be historical, i.e. the time passed since the institution of the Littoral was too short to effectively reorganize the various subregional entities. Indeed, the Littoral was a mosaic of territories that had joined the Hapsburg empire in different times and circumstances, coming from very diverse kinds of politi—the latest being the former Venetian lands. However, 60 years—possibly 100 years considering the French period—is a lot of time. It is necessary to look further into the larger context of the territory: first and foremost, the structure of the Hapsburg Empire. Until its dissolution after WW1, it had maintained the organization of an *ancien régime* state, based on a semi-feudal system, which worked well in a multi-ethnic and multireligious context. In the specific case of the Littoral, its subcomponents had a long history of conflicts; moreover, the development of the coastal areas had been blocked and re-oriented for centuries to serve the interests of the maritime and commercial Venetian state.

[11] Actually the first draft of the *regioni* dates back to the statistical districts of the kingdom of Italy established in 1864 (Treves 2004; Lando 2011), that the highly centralized Italian government made somehow official by adopting in 1912 the term "regions". Effectively this was a recognition of the cultural turn carried out at the time by French geographers. These operate, instead, rediscovering the individual regional identities through a program of scientific research. After WW2, in 1946 Italy's the new republican government decided to attribute political functions to these spatial units, in an early attempt to decentralize its power structure.

Indeed, the opposition between Venetian and non-Venetian areas was a crucial, and oftentimes overlooked, factor in the territorial and socioeconomic organization of an area much broader than the Littoral, covering the whole eastern coast of the Adriatic Sea. The Venetian state was different from its neighbors, all land-based powers. Venice had a structure closer to ancient Mediterranean city-states, such as Carthage and Athens than modern maritime powers, such as England. Venice's power relied on a tributary region centered on the sea, supported by a network of military bases and administrative outposts located along the main trade route linking Venice with the Middle East. Thus, the backbone of Venice's urban system consisted of several cities along the eastern coast of the Adriatic Sea, corresponding to today's Croatia and Albania. Such territorial arrangement was Venice's only option because the Ottoman Empire prevented it from expanding inland into the Balkans, and a coalition of Western powers limited its conquests in the Italian Peninsula.

The geopolitical struggles between Venice and its opponents, compounded by the physical geography of the region featuring a mountain range parallel to the coastline (the Dinaric Alps), contributed more than the ethnolinguistic differences to the establishment of a dual territorial system. On the one hand, the interior areas were a relatively closed system, with its own culture, autarkic economic system, and locally focused political system; on the other hand, the coast was an open system, connected with Venice, ruled by an Italian-speaking elite.

After two world wars, ethnic cleansing and forced migrations canceled, sometimes bloodily, some ethnic distinctions, but did not erase the two territorial systems, mostly because movements of people depended on a transportation infrastructure whose basic design dated back to the Napoleonic period. The situation has begun to change only in the last two decades, thanks to the dissolution of Yugoslavia and the formation of independent nation-states in Slovenia, Croatia, Bosnia-Herzegovina, and Montenegro. Each new government has worked to integrate coastlines and interiors, breaking the latter's isolation by building national transportation networks to the Adriatic Sea (Battisti 2000).

In 2001, the census portrayed the outcome of 60 years of evolution since the partition of the former Littoral between two and then three different cultural, political, and economic entities. Indeed, since then there have multiple collaborations across the border (Battisti 1986), and several state-led initiatives to promote the functional integration in the former Littoral (Battisti 2011). However, it is an illusion to expect that these efforts could do anything more than delaying an unavoidable divergence, especially because the Littoral has never been functionally united.

References

Battisti G (1979) Una regione per Trieste. Studio di geografia politica ed economica, Trieste, Ist. di Geografia, Fac. Econ. e Comm, Università degli Studi

Battisti G (1986) «Le relazioni frontaliere sul confine tra Friuli-Venezia Giulia e Slovenia 1982–1985. Tra crisi ed evoluzione». In: Atti 10° Convegno Scientifico Internazionale Alpe-Adria (Trieste, 24–25 ottobre 1986), Facoltà di Economia e Commercio, Università di Trieste, pp 61–103

Battisti G (1993) Trieste e il suo porto. I modelli di riferimento di un'evoluzione urbana. In: Panorama di Trieste attraverso vedute e piante della collezione Davia, a cura di Cammarata M, Trieste, Ed. La Mongolfiera, pp 86–117

Battisti G (2000) «La Dalmazia. Natura e funzioni di un organismo geografico». In: Contributi alla conoscenza della Storia e della cultura dell'Istria, di Fiume e della Dalmazia, a cura di S. Cattalini, Udine, ANVGD, pp 31–47

Battisti G (2011) Gorizia Nova, un'eurocittà al confine tra Italia e Slovenia. In: A Pasquale Coppola. Raccolta di scritti, Memorie della Soc. Geogr. Ital., LXXIX, a cura di L: Viganoni, Roma, vol I, pp 43–56

Battisti G (2014) Equilibri in mutamento nell'Europa del terzo millennio. Il caso del confine orientale d'Italia. In: Atti XXI Congresso Geografico Italiano, a cura di G. Scaramellini e E. Mastropietro, Milano, Mimesis, vol I, pp 277–287

Capineri C, Celata F, de Vincenzo D, Dini F, Randelli F, Romei P (eds) (2014) Oltre la globalizzazione. Resilienza/Resilience, Firenze, Società di Studi Geografici

Cervani G (1980) Maria Teresa e Trieste. Il Litorale austriaco come provincia storica. Quaderni giuliani di storia, 1(2):33–54

Christaller W (1933) Die zentralen Orten in Süddeutsch-Land. Gustav Fischer, Jena

Cori B, Cortesi G, Formentini U (1976) Studi su città, sistemi metropolitani,sviluppo regionale, II Quaderno: Studi sulla rank-size rule, a cura della Commissione di ricerca del Comitato dei geografi italiani, Pisa, Ist. di Geografia economica dell'Università di Bologna

Cori B (1976) Rank-size rule e armatura urbana dell'Italia. In:Studi sulla rank-size rule, Pisa, Ist. di Geografia economica dell'Università di Bologna, pp 17–26

Di Napoli MF (2014) L'invenzione della regione. La soggettività della regionalizzazione e il caso della Brianza, Tesi dottorale in Scienze dei Beni Culturali e Ambientali, Università di Milano, XXVII Ciclo, a.a. 2013–14

Fonseca JW (1988) Urban rank-size hierarchy. A mathematical explanation, Fairfax (VA), George Mason University, WP no 8

Galluccio F (1998) Il ritaglio impossibile. Lettura storico-geografica delle variazioni territoriali del Lazio dal 1871 al 1991, Roma, DEI

Glaser EL, Kolko J, Saiz A (2001) Consumer city. J Econ Geogr 1(1):27–50

Groenewegen P (2013) "Introduction", Alfred Marshall, Principles of economics, 8th edn. Palgrave Macmillan, Basingstoke, Palgrave Macmillan, XIV

http://pxweb.stat.Si/pxweb

http://www.dzs.hr/.Hrv.censuses/census2001.htm

ISTAT (2001) 14° Censimento generale della popolazione e delle abitazioni, Roma

Jefferson M (1939) The law of the primate city. Geograph Rev 29:226–232

Lando F (2011) Le regioni da Piero Maestri alla Costituzione. In: Muscarà C, Scaramellini G, Talia I (eds) Tante Italie Una Italia. Dinamiche territoriali e identitarie. Vol. I: Nodi e modi della nuova geografia, Milano, Franco Angeli, vol I, pp 13–40

Marshall A (1890) Principles of economics. Macmillan, London

Michieli R., G.Zelco (a cura di), (2008) Venezia Giulia. La regione inventata, Udine, Kappa Vu

Mogavero V (2016) L'invenzione' del Veneto tra identità sospese, frammentate e multiple. https://storiaeregione.eu/it/news-eventi/leggi-evento/colloqui-bolzanini-di-storia-regionale

Muscarà C (1968) Una regione per il programma, Venezia, Marsilio

Pivec–Stelé M (1930) La vie économique des provinces illyriennes (1809–1813) suivie d'une bibliographie critique, Paris, Inst. d'Etudes Slaves de l'Université

Raffestin C (1984) Territorializzazione, deterritorializzazione, riterritorializzazione e informazione. In: Regione e regionalizzazione, Milano, F. Angeli, pp 69–82

Rossi D (2020) L'invenzione di una regione. Le radici storiche dell'autonomia in Friuli, Venezia Giulia, Istria Fiume e Dalmazia nel lungo Novecento [The invention of a Region. The hisorical roots of autonomy in Friuli, Venezia Giulia, Istria, Fiume and Dalmatia in the long twentieh century], Historia et Ius, pp 1–22

Salimbeni F (1991a) Graziadio Isaia Ascoli e la Venezia Giulia. In: Dal Litorale austriaco alla Venezia Giulia. Miscellanea di studi giuliani, Udine, Del Bianco Editore, pp 51–68

Salimbeni F (1991b) AA VV, Dal Litorale austriaco alla Venezia Giulia. Miscellanea di studi giuliani, Udine, Del Bianco Editore

Schiffrer C (1953) I centri slavi degli altipiani carsici triestini e la loro evoluzione ad opera degli Italiani. Boll Soc Geogr Ital, pp 453–470

Spezialorstrepertorium der österreichischen Ländern (1918) Bearbeiten auf Grund der Ergebnisse der Voelkszaehlung vom 31 Dezember 1910, VII Oest-Illyr. Kuestenland, Wien

Tong Soo K (2004) Zipf's law for cities: a cross country investigation, London, L.S.E., CEP Discussion Paper no 641

Treves A (2004) I confine non pensati: un aspetto della questione regionale in Italia. Annali della Facoltà di Lettere e filosofia, Università degli Studi di Milano, LVII(2):243–264

Turco A (1984) Lo spazio non-regionalizzato: una versione sistemica. In: Turco A (ed) Regione e regionalizzazione, Milano, F. Angeli, pp 83–104

Valussi G (1971) Il riassetto amministrativo ed economico dei territori della Venezia Giulia annessi alla Repubblica Socialista Slovena. In: Atti del XX Congr. Geogr. Ital. (Roma, 1967), Roma, Soc. Geogr. Ital., vol IV, pp 272–294

Valussi G (1972) Il confine nordorientale d'Italia, Gorizia, ISIG

Venezia Giulia (1946) study of a map of the Italo-Yugoslav National Borders, Rome, C. Colombo

Verani F (1990) La questione etnica ai confini orientali d'Italia. Antologia a cura di F. Verani, Trieste, Edizioni "Italo Svevo"

6 *Gorizia Nova*, Aka "New Gorizia:" A Euro-City on the Border Between Italy and Slovenia. A Recommendation for Local-Level Territorial Changes After Slovenia Joined the European Union in 2004

Abstract

While the previous chapter discussed the micro-level territorial impact of two world wars on the region surrounding Trieste, this chapter delves into the possibilities of reterritorialization following the enlargement of the European Union (EU) to include neighboring Slovenia. It focuses on a settlement north of Trieste, whose urban area was split after WW2 between the Italian town of Gorizia and the Slovenian new town of Nova Gorica. EU membership of both Italy and Slovenia led to the removal of border checkpoints, opening the possibility to re-integrate the urban space of the two towns. However, the two countries used vastly different criteria to draw the boundaries of their towns and rural areas, which constitutes an obstacle to joint territorial planning. Using historical geography and GIS techniques, this chapter explores these different territorial logics and proposes a redrawing of administrative boundaries for the purpose of joint urban planning, which is also respectful of both countries' existing administrative units.

Keywords

European Union enlargement · Italy · Slovenia · Re-territorialization · Gorizia · Urban geography · GIS

6.1 Introduction: Integrating the Two Gorizias

The first chapter of this section dedicated to the historical geographies of the *Triestino* region have focused on the entanglement between the 'grand,' realist geopolitics of the nineteenth and twentieth centuries and the micro-level implications on the territories surrounding Trieste. At the bottom line, Battisti's goal was to show the resilience of administrative infrastructures, economies, and peoples when faced with the consequences of world-shattering global conflicts. This chapter, based on research conducted in the Department of Geography—*Dipartimento di Scienze Geografiche e Storiche*—of the University of Trieste in the mid-2000s deals with the consequences of world-changing, continent-wide peace.

In 2004, Slovenia joined the European Union (EU); in 2007, it adopted the European single currency, the Euro; and in early 2008 it joined the Schengen Area, thus removing border checkpoints with other EU member states, including Italy. For Trieste and the surrounding areas, such changes meant an opportunity to overcome the border, and to reconstitute the social and economic/commercial space of a portion of the old Littoral. This had been transformed in 1918 with the annexation to Italy, broken in 1945 due to the territorial gains of Yugoslavia at Italy's expenses, and further

C. Sellar and G. Battisti, *Geopolitical Perspectives from the Italian Border*, Historical Geography and Geosciences, https://doi.org/10.1007/978-3-031-26044-5_6

partitioned in 1991 with the breakup of Yugoslavia and the independence of Croatia and Slovenia. However, free movement of people and single EU currency did not mean an automatic reintegration of regional economies and urban infrastructures, especially in a territory that, as shown in the earlier chapter, always had a questionable level of integration.

Urban reintegration is an especially visible issue around Gorizia, Italy's smallest provincial capital, a province whose territory borders that of Trieste. During the Cold War, Gorizia was one of several divided cities of Europe—the most famous of which was Berlin. Territorial shifts after WW2 and the ideological divisions between East and West had brought international borders to cut across some urban areas. In Gorizia, the border change broke apart a territory that had been politically and economically united for centuries, first as a principality within the Holy Roman and then Austrian empires, and later as a province of Italy. Like Berlin, Gorizia sported its own (much shorter) wall. Unlike Berlin, the new border cut the outskirts of the city: the center remained in Italy, and a few peripheral neighborhoods became Yugoslav. Such territorial unbalance prompted the Yugoslav government to build a new urban center, called *'Nova Gorica.'* The new city duplicated Gorizia's urban functions; thus, both cities potentially relied on roughly half of the hinterland of the former principality. With the dissolution of Yugoslavia, the newly independent Slovenia inherited Nova Gorica and its divided city status.

In the new millennium, Slovenia's membership of the EU, the Euro, and Schengen opened up the opportunity to rebuild the economic space of the old Gorizia; if not as an administrative unit, at least in terms of a functionally integrated urban area. However, such reintegration required, and still requires, a lot more than tearing down some walls and checkpoints. It requires the harmonization of administrative units built following profoundly different logics, the elimination of duplicate urban functions (airports, hospitals, and the like), and integration of infrastructures (roads, sewage, phone and Internet lines, and others). Figure 6.1 shows the territories surrounding the two Gorizias: in Italy, Gorizia is the seat of government of a province bearing the same name; in Slovenia, Nova Gorica is the center of a large statistical region (Goriška statistična regija, est. 2000). Just looking at the relative sizes of the two territorial units

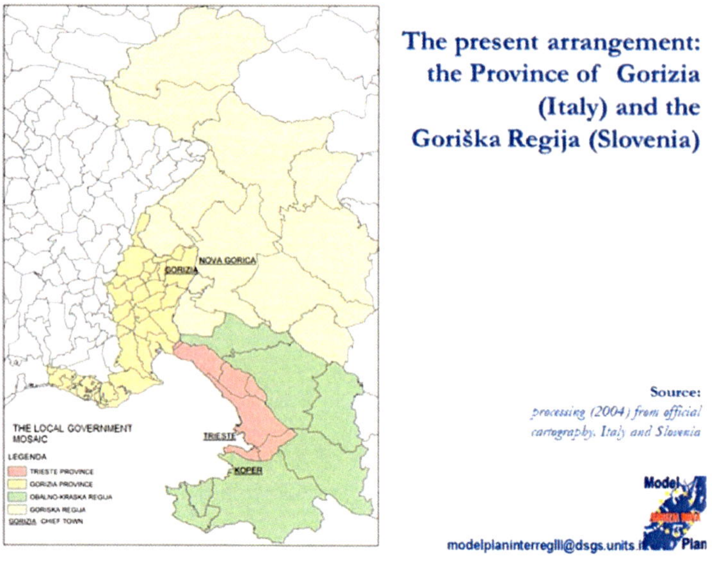

Fig. 6.1 Gorizia area: local governments on the Italian-Slovenian border. *Source* ModelplaninterregIIIa

highlights the diversity of administrative structures and logic in the two countries. The title of this chapter, Gorizia Nova, aka 'New Gorizia,' identifies such a yet-to-be-realized project of territorial reintegration. The name Gorizia Nova emphasizes inclusiveness of the two major ethno-linguistic communities living in the area, and a nod of respect to the long history of the city and region: Gorizia is the city's Italian name, Nova means *new* in Slovenian and also signifies the inclusion of the community of Nova Gorica. Moreover, it also means *new* in Latin, the language of high culture throughout the middle ages and the early modern period. As such, it signifies the legitimacy deriving from the one thousand years-old existence of the city. It also aims at avoiding susceptibilities from both countries, in a human environment where the names have often been used as political weapons. The remaining sections of this chapter identify, first, the main challenges to reintegration and, second, the territorial administrative structures best suited to implement reintegration policies.

6.2 The Territorial Challenges of Gorizia Nova

On both the Slovenian and Italian sides of the border, local administrators in the twin provinces of Gorizia and Goriska complain of being marginalized by their respective states. Local leaders in both provinces have often struggled—within their respective states—to defend their autonomy against the neighboring provinces that were threatening to absorb them. From their perspective, the fall of the border is an opportunity, because the restoration of territorial continuity could lead to the birth of a unitary urban area with double the population and resources of either city—too big and wealthy to be marginalized/annexed by others. Simply put, a functional reintegration would bring the two provinces to the status Gorizia had before WW2, and that the postwar institution of a dense network of border crossings (21 international, local, and "agricultural" crossings) was not able to maintain.

Achieving functional unity will require the political solution of delicate problems of an economic, cultural, and urban nature. Given the role of territorial coordination inherently belonging to an urban organism located in the basin north of the confluence of the two main rivers of the area, the Vipava (in Italian: Vipacco) and the Isonzo (Slovenian: Soca), the successful solution of contingent political problems will depend on its management. In a nutshell, a successful Gorizia Nova will have to play a political and economic role for a specific area, bounded by physical geographical features and characterized by similar economic activities. Challenges notwithstanding, Gorizia Nova has good chances to succeed. First, the relative demographic balance on the two sides of the border means that one city/province will not absorb the other, thus creating an opportunity for local administrators to collaborate in equal terms. The two Gorizia provinces host a roughly equivalent population (120,000 in Goriska, 130,000 in Italy), with a wide disparity in surface area (2325 km^2 in Slovenia, just 470 in Italy). Second, the combined urban population is large enough to guarantee political weight both in Italy and Slovenia. Together, the urban areas of the two provincial capitals include 73,000 inhabitants (2002 data), equally distributed between the two entities, although with a very different density. Such population means that Gorizia Nova would be the third-largest city of its region in Italy, very close to the second (Udine), and also the third largest of the whole country of Slovenia, also in this case not very far from the second (Maribor).

The coexistence of two similarly sized neighboring cities and provinces on either side of the border has facilitated the establishment of positive, collaborative relations between the populations and local authorities across the border. Most notably, the two provinces signed a "cross-border pact", which has enabled an operational dialogue between the technical bodies responsible for the management of public services. Provincial and urban authorities have been willing to co-operate for a long time, at least since the peace treaty between Italy and Yugoslavia in 1976. They were aware that the divisions caused by the Cold War were not tenable in the long term, and that the area of the two Gorizias was likely to develop into a cross-border collaboration similar to many others across Europe.

Collaborations notwithstanding, further integration between the two areas will require deep reforms, because the administrative structures of the two Gorizias are profoundly different, not only with concern to the languages spoken on the two sides of the border (Italian and Slovenian, with minority areas on both sides) and the organization of the townships. There are also the roles of the two administrations in the local economies, which are different too: purely urban on the Italian side, and mixed urban and rural on the Slovenian side, because the Slovenian township includes a larger territory encompassing Nova Gorica's countryside. Only if the two cities and provinces will decide to establish a joint territorial planning commission there will be hope to harmonize the territories and the economies of the two areas, which is the necessary precondition for integration.

To even consider the option of a joint planning commission, there is a need to solve a series of technical and scientific issues, starting with the harmonization of cartography. Thus far, the different logics of administrations on the two sides of the border led to the production of substantially different maps, an underlying different conception of the role of local governments in their respective territories. Simply put, there is the need to reach an agreement over the appropriate territorial boundaries of the two entities to be integrated. To lay the groundwork for such integration, the Department of Geography—*Dipartimento di Scienze Geografiche e Storiche*—of the University of Trieste (at the time chaired by Battisti) was awarded a European Union grant for a project titled 'Gorizia Nova Model Plan. Models for sustainable development of the border areas.'[1] The remainder of this paper discusses the project's findings concerning cartography.

6.3 The Identification of the Areas

Any analysis of the areas surrounding Gorizia and Nova Gorica requires a cartographic identification of the two cities, which is not easy to accomplish, because, as the previous paragraph showed, administrators used different criteria to draw their boundaries.[2] The Italians drew city boundaries based on urban density, while the Slovenians used the criteria of functional integration among spatially distinct areas. Here opens the unresolved question of the polymorphic nature of the urban concept, which has found in the literature a plurality of solutions, each one functional to a given setting. On the Italian side, cartographers tended to privilege the material fact of the continuity of the built-up area; on the Slovenian side, they relied on the relationships that unite settlements detached from each other by large swaths of agricultural and forested areas.

By simply adding up the two existing municipalities facing each other along the border, one would obtain an apparent balance in terms of population (about 35,000 inhabitants on

[1] Community Initiative Programa Interreg IIIA, Italia-Slovenia 2000–2006, Gorizia Nova - Model Plan. Models for the Sustainable development of border areas. Project cofinanced by the European Union through the ERDF within the framework of Interreg IIIA Italy-Slovenia under the direction of the Autonomous Region Friuli Venezia Giulia. The bibliography only reports the conclusive summary, deprived of the very rich cartographic equipment (Dept. of Geography, University of Trieste 2006).

[2] Moreover, the two countries do not make use of the same cartographic base. Both choose inverse cylindrical projections, but while in Italy the official cartography adopted the Gauss–Boaga projection and (especially in the digital versions) also the UTM system, in Slovenia they use the Gauss-Krueger one. The missing link between the two cartographies is particularly disturbing at the scale required by spatial planning. So a preliminary standardization is needed (Favretto and Mastronunzio 2006).

both sides according to the latest censuses, which could be advantageous from a political point of view). The result, as can be appreciated from the map, would be very unbalanced under the profile of the territorial size and spatial structure of the settlement and related economic activities. Such choice would be a double mistake. In addition to uniting the urban and rural areas (as can be deduced from the average settlement density of 215 inhabitants per square kilometer) which would cause urban planning problems, later on, it would also exclude extensive urban areas belonging to other smaller municipalities located along the river, which would instead benefit from joint infrastructural planning. If the latter were included, this would lead to a maxi-area of 294.3 km^2, with almost 83,000 inhabitants. This is a less exclusionary entity, but consequently even more spatially unbalanced, and quite frankly too big to manage with the given population and tax base.

A dimensional rebalancing in terms of the surface area appears to be of difficult solution. The municipality of Nova Gorica covers 309 km^2, more than seven times that of Gorizia (41.11 km^2). Wanting to find an administrative area of comparable size in Italian territory, it would be necessary to refer to the whole province. The latter, in fact, with its 466.02 km^2, is 50% larger than Nova Gorica and has a population four times that size. Given the different administrative statuses of province versus municipality and the larger population and territory of the Italian province, such a proposal would send the wrong political message, equivalent to an intent of annexation, which is unacceptable given the tormented history of the area. The only merit of such a proposal would be the plurality of settlements—47 in the municipality of Nova Gorica, versus the 25 municipalities of the Italian province. Thus, it is necessary to go back to the map and try to carve out comparable areas on either side of the border.

One possible solution consists in looking at the districting of public services. Nova Gorica performs functions of coordination and provision of public services for the Slovenian province, including all 46 settlements in the municipality. On the Italian side, it is necessary to identify a district for the provision of certain public services that covers an area intermediate between the township of Gorizia and the Province. A possible candidate is the service area of the Employment Center of Gorizia, which coincides with the *Alto Isontino* District of the healthcare services provider *Azienda per i Servizi Sanitari n. 2*. It consists of 16 municipalities with 71,011 inhabitants (as of the end of 2004) and 219.18 km^2. If we add the municipality of Doberdò, we get the territorial scope of the School District of Gorizia, for a total of 240.03 km^2 and 72,469 inhabitants (Census 2001, 2002). This is a surface area comparable to Nova Gorica, but with more than twice as much population. From the point of view of the border areas most directly influenced by Slovenia's membership of the European Union, these two areas can constitute the core of a unified transborder district, thanks in no small measure to physical geography, because they both gravitate on the valley of the *Isonzo/Soca* River, which creates opportunities for interconnected infrastructures and the rationalized provision of public services. Using the *Isonzo/Soca* Valley as criteria of agglomeration would require to include three additional Slovenian townships—Kanal, Smartno, and Miren-Kostanjevica. The outcome would be a relatively balanced area, centered on the urban core of the two Gorizias.

6.4 Defining *Gorizia Nova*'s Urban Core

Wanting to proceed to a first cartographic definition of an urban core encompassing both Gorizias, we opted for a practical but imprecise solution. Rather than trying to propose an unquestionable solution to a fundamentally political problem, the goal is to start a conversation involving both the academic community and policymakers. The analysis of the density of buildings at the neighborhood level shows a contiguous, high-density area covering most of the Italian municipality and roughly 70% of the settlements constituting Nova Gorica. Beyond such core, there is a historical legacy of dispersed small settlements, driven by geomorphology: hills and steep valleys separating long-established rural communities, some of which had opportunities to grow into small urban centers at various points in time. In defining the core, the author arbitrarily chose a population density of 351 inhabitants per square kilometer, which is an unquestionably urban density. Such threshold was established by lowering the criteria established in the 1970s for the analysis of the urban structure of Lombardy, because since then the urban lifestyle has spread to rural areas, and many European cities have developed suburban neighborhoods (Battisti 1977).

The result is a relatively compact urban center encompassing both sides of the border in a Southwest/Northeast direction, surrounded by a transitional belt of rural and suburban areas. Beyond such a belt, there is a crown of smaller urban areas, such as Gradisca in Italy, Miren, Dobrovo, Kanal, and Dornberk in Slovenia. These towns are the external limits of a territorial system that preserves a large share of nature within it, which allows for a "buffer zone" for a future possible expansion of the urban core. The latter includes mature, high-density urban areas, with some neighborhoods on both sides of the border exceeding 1500 inhabitants per square kilometer.

Such distribution of the population shows that the hypothetical Gorizia Nova already has the characteristics of a single city, not simply of an urbanized area consisting of functionally separate towns that are absorbing the rural areas that separate them. Gorizia Nova has the typical structure of a monocentric agglomeration, in which the heart of the system is—seemingly paradoxically—constituted by the borderline. The sharp growth of density toward the center is characteristic of the cities in a phase of growth, a very different scenario from the urban decline of the other major cities of the area, especially Trieste and Udine in Italy. Such regional shift of population away from the major cities toward small- and medium-sized towns is consistent with similar patterns throughout the industrialized world. Smaller urban areas are growing, due to their higher quality of life and the reserve of productive factors, including spaces available for building, the availability of human and financial resources, and the greater efficiency of services, which makes them privileged destinations for both labor and capital. Such conditions bode well for the future of Gorizia Nova.

Opportunities notwithstanding, the process of establishing, or even just agreeing to, an administrative, joint-planning commission will not be easy. Using the aforementioned threshold of 351 inhabitants per square kilometers, the proposed boundary will involve at least four municipal governments: Gorizia and Sempeter-Vrtojba for their whole territory; and Nova Gorica and Miren-Kostanjevica for only a part of their territory. Such a project will require strong commitment and buy-in from the leadership and administration of all the local governments involved. Such buy-in will not be easy to achieve, due to the balance between Italian and Slovenian administrators, and the request to involve parts of existing municipalities. For example, Sempeter-Vrtojba had been part of the municipality of Nova Gorica, and the local administrators may or may not support what they will likely perceive as a re-annexation of their town to the larger city. Furthermore, following EU regulations both Italy and Slovenia are devolving some of their central governments' power to regional and city governments, which will require the involvement of neighboring municipalities in the Gorizia Nova project. For example, on the Italian side, the township of Savogna shares with Gorizia the management of the airport and the industrial park—which will be important focuses for the proposed joint planning commission. If these delicate political problems will be resolved, the territorial complex thus identified will count 71,311 inhabitants (of which 33.922 in Slovenia), distributed over a surface of 13.477 ha (of which 8.457 in Slovenia), for a density of 529 inhabitants per square km in Italy and 401.1 in Slovenia. Figure 6.2 shows what this proposal looks like on the territory: a spatially and demographically balanced urban center (identified as 'Gorizia Nova Perimeter' on the map) with a clear hinterland (Service area on the map) cutting across existing administrative units of various nature in both countries.

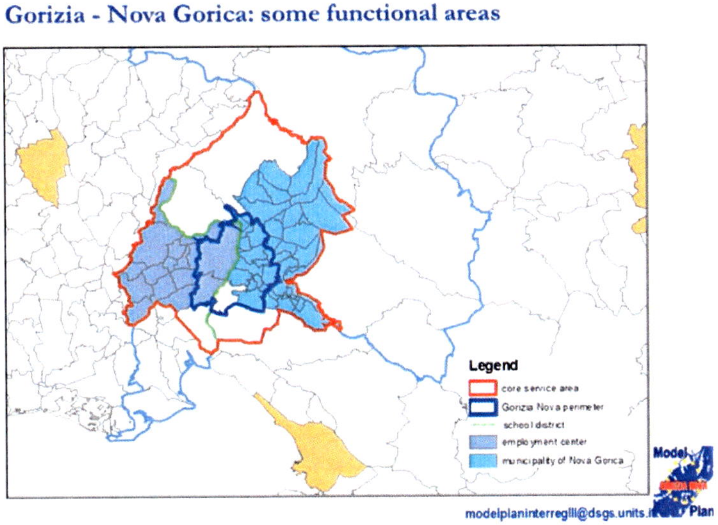

Fig. 6.2 GoriziaNova functional areas. *Source* ModelplaninterregIIIa

If the proposed boundaries, shown in Fig. 6.3, were to be accepted by the interested layers of governments, the resulting jointly planned territory would allow: (a) to include the whole area with a certain urban density; (b) to involve the whole transnational belt; (c) to leave adequate expansion spaces to the agglomeration, both in Italian and Slovenian territory; (d) to solve the problem of the double planning areas in the Italian territory, guaranteeing, in particular, the common management of the industrial area and the airport site, located across the municipalities of Gorizia and Savogna d'Isonzo; (e) to reconnect all the administrative units that have left the municipality of Nova Gorica; (f) to involve an acceptable portion of the municipality of Nova Gorica (3. 415 ha out of 11.09). Of course, this is just a working hypothesis; the Autonomous Region Friuli-Venezia Giulia and the Municipality of Nova Gorica would have to approve it and send it to the attention of the European Union authorities.

The main objection the author expects from the political authorities is that the proposed unification implies in some way a separation—albeit limited to certain practical aspects of urban government—within the municipal territory of Nova Gorica. Leaving aside considerations of a political nature, it should be noted that the hypothesis is part of the trend that emerged in Slovenia after independence, toward the multiplication of municipal autonomies, breaking up the large municipalities inherited from the socialist system. The latter had continued along the line opened by the Italian administration, which had repeatedly unified the micro-municipalities in which the Austrian administrative network was articulated. A rationalization that was certainly positive from a technical point of view between the two world wars, but less so after 1947, which led to the birth of very large municipal districts.

The choice to establish 'mammoth municipalities' was consistent with the tendency to centralize the functions of coordination and control, typical of socialist regimes pursuing planned economies (Geograph Slovenica 2000). In the wider region of the old Austrian Littoral, the unification was functional to the development needs of the urban framework. After the dissolution of Yugoslavia in 1991 and the affirmation of democratic governments and market

Fig. 6.3 Gorizia Nova: a tentative framing. *Source* ModelplaninterregIIIa

economies, the need for attention to the problems of citizens in the context of an economy based on a private initiative led to a new phase of deterritorialization and reterritorialization. Simply put, Slovenian decision-makers realized that democratic participation and private business benefited from the more direct access to government enabled by smaller municipalities. In fact in the larger territories of the socialist era local governments had merged villages and small towns often times separated by historical legacies and difficult to cross mountain areas. In this specific case, the current trend seems to prefigure a gradual process of dismantling of the large municipality of Nova Gorica, whose replacement with a plurality of local authorities would necessarily open the way to a redefinition of the power to manage the city. The perspective at this point, however, should be cross-border. As the mayor of Nova Gorica, Mr. Brulz, has officially underlined (19 April 2006): "The creation of the Goriska region is important for us. But it is even more important to obtain the recognition of the Gorizia Euroregion, to give this territory an additional instrument of development on both sides of the border."

6.5 Conclusion

In writing this chapter, we showed that the peace that followed the enlargement of the European Union was not free of challenges for the areas surrounding Trieste. In the nearby 'divided city' of Gorizia, Slovenia's EU membership opened up tremendous opportunities to re-integrate the economies and infrastructures of communities with a long history of political unity that was split up in 1947. However, this paper's technical discussion about cartography and appropriate administrative boundaries shows, in a nutshell, that the ideological division between East and West was not only a matter of 'grand' geopolitics. It was also a matter of profoundly different logics of producing territories, especially local-level administrative subdivisions, led by different goals of how and why municipalities and provinces should provide services to their citizens. Overcoming these differences is hard work, including both political will and technical expertise. Among experts, cartographers can play a key role, provided they are sensible to history, urban structures, economies, and to the physical geographical features of the environment. Specifically, the Gorizia Nova project is meaningful because of the high urban density along the border, the long shared history of the old principality, and its location. The latter is especially relevant because the Isonzo/Soca river valley makes joint infrastructural planning rational and relatively easy to execute. Once more, 'Battisti's' geography aims at being an integrative discipline, where the history of places, their environments, politics, and economies converge. With such notion of the discipline in mind, the following chapters turn to higher geographical scales, considering the geopolitics underpinning European and global trends.

References

Battisti G (1977) Contributo alla delimitazione territoriale della « regione-citta» milanese [Contribution to the territorial bounding of Milan's 'city-region']. In: Corna Pellegrini G (ed) Milano Megalopoli Padana Valli Alpine. Studi sulle reti urbane, Bologna, Patron, pp 179–216

Data sources: census, 2001 and 2002

Dept. of Geography, University of Trieste (2006) Gorizia Nova. The Urban Area Gorizia—Nova Gorica as a Model for Europe. Final Report, Trieste, Regione autonoma Friuli Venezia Giulia. http://www.theslovenian.com/articles/gorizianova.pdf

Favretto A, Mastronunzio M (2006) Integrazione delle cartografie nazionali per le aree transfrontaliere. Un caso applicativo: l'area urbana di Gorizia-Nova Gorica [Connecting maps with different coordinate reference system in a digital format. An application in the urban area of Gorizia (Italy) and Nova Gorica (Slovenia)]. In: Bollettino A.I.C. n. 126–127–128, pp 307–316

Regionalni razvoj v (2000) Sloveniii [Regional development in slovenia] special issue. Geograph Slovenica 33(2)

Check for updates

7 Inland Areas and Border Regions: A Geopolitical Interpretation. Comparing the Marginalization of Trieste and Umbria as Examples of the Dynamics of Borderlands Versus Remote Inland Areas

Abstract

This chapter concludes the book's section dedicated to producing geopolitics from the vantage point of Trieste, drawing on historical geography, it tackles another relevant issue in contemporary human geography: the marginalization of places and communities. It compares the border regions of Trieste and Bolzano—both former Austrian lands on Italy's border—with the landlocked region of Umbria, in the center of the Italian peninsula. In so doing, it argues that the historical geographies of borders have long-lasting effects on marginalization, which may last even centuries after the borders have moved or ceased to exist. This is especially the case in areas not touched by the arbitrary border-making of colonialism: old countries and empires tended to set their borders in sparsely populated or difficult-to-traverse areas. Therefore, physical geographies often compounded the political and infrastructural divisions created by the border. This is especially the case in Umbria, a border area between various Italian states in the medieval and modern periods, until the Italian unification in 1861. On the other hand, new borders breaking previously existing territorial units (such as the areas around Trieste and Bolzano) may damage economic integration with long-lasting effects.

Keywords

Core-periphery underdevelopment · Inland areas · Borderlands · Geopolitics · Historical geography

7.1 Introduction: The Marginality of Internal Areas

The question of internal areas arises from the problem of economic development, which recognizes two types of situations: development and underdevelopment. This viewpoint, which is rooted in 19th-century positivist science and evolutionary approaches, builds on the axiom that there is a single "path of development" that all territorial realities must sooner or later follow (Rostow 1960). Such approach considers underdevelopment as an equivalent of delayed development, which is the result of either a disadvantaged starting condition or an insufficient degree of positive modification over time compared to the conditions of the "leading" areas. Built-in this definition is the expectation that underdeveloped areas should sooner or later move on to a condition of "development." In practice, since the highly advanced areas are, by definition, limited in numbers, most observers implicitly recognize three situations: "normal" development, underdevelopment, and "overdevelopment". In

C. Sellar and G. Battisti, *Geopolitical Perspectives from the Italian Border*,
Historical Geography and Geosciences, https://doi.org/10.1007/978-3-031-26044-5_7

quantitative terms, this can be easily defined with an average value—in relation, for example, to the gross domestic product or one of the various indicators in use today (human development index, etc.)—and the situations significantly below or above this average.

7.2 Geographical Marginality and Economic Marginality

From this point of view, inland areas can rightly be defined as *marginal*[1]—a term akin to the "existential peripheries" of which Pope Francis speaks.[2] Such definition of marginality (or peripherality) implies a relationship with a center whose mechanisms of accumulation depend on centripetal flows of productive factors, extracted from the peripheries. As historically verified, geographical marginality is a condition that in itself favors the perpetuation of situations of underdevelopment. Far from the pulsating centers of the economy, it is more difficult to set in motion virtuous cycles of economic growth, which increasingly require the influx of productive factors from outside. This is all the more difficult because it requires contrasting an established outflow of local productive factors toward the center. In this chapter, the notion of marginality enables to subsume under the same analytical category the internal marginal areas and the "external" (or peripheral) marginal areas, i.e. the borderlands.

If we look at the physical reality of borderlands, we are reminded of how in many instances, especially in long-established states in Europe and Asia that did not experience the arbitrary border-drawing of the colonial period, the boundary lines are usually supported by morphological elements that constitute, rather than points of reference, barriers to free movement. Whether they are reliefs or bodies of water, swamps, or deserts, most of these borderlands occur in marginal areas, where the scarcity of population is the product of a scarcity of resources, if not of an environment hostile to large human settlements. An example of these geography-driven historical borderlands is the Pontine Marshes that marked the border between the Papal States and the Kingdom of Naples from the Middle Age to the Italian unification in 1861. Before the great land reclamation of the 1930s, these marshes were difficult to traverse and carried a high risk of malaria, thus preventing the movement of invading armies between the two states.

Along the border, two (if not more) contiguous areas end up "facing each other", which are, by definition, "marginal" to the rest of the respective national territory. In concrete terms, the two opposing fronts tend to share similarities even when communication between the two parties is impeded. When we then look at the reality of the places we almost always end up discovering that in the long run the border does not so much produce a "scar", that is, a line that crosses the territory, because the condition of marginality on both sides tends to recompose in some way an environment that the border fact has tried to divide, with different consequences from place to place. Figure 7.1 represents such tension, built in border areas: on the one hand, the representation of a sharp dividing line, characteristic of the geographical imagination of the modern nation-state; on the other hand, the lived experience of the borderlands, much more akin to transboundary, transition zones. In the perspective of spatial planning in Europe, where the European Union has emphasized collaboration across the borders of the member states, the result is usually the creation of a larger marginal

[1] Between 1980 and 1986, about fifty researchers from 21 university sites turned their attention to this issue, under the auspices of the *Gruppo Rivalorizzazione Aree Marginali* [Group for the Revitalization of Marginal Areas] (GRAM), of the Association of Italian Geographers (AgeI) (Cencini et al. 1983; Leone 1986, 1988). A National Strategy for Internal Areas will be launched in 2014 (Sommella 2017).

[2] Existential peripheries are "the places where there is suffering" (homily of the Chrism Mass, 28/3/2013) or rather "the places that are inhabited by all those who are marked by physical and intellectual poverty". To them was dedicated n. 20 of the Apostolic Exhortation *Evangelii Gaudium*, [Gospel of Joy] where they are defined as "the peripheries that need the Gospel".

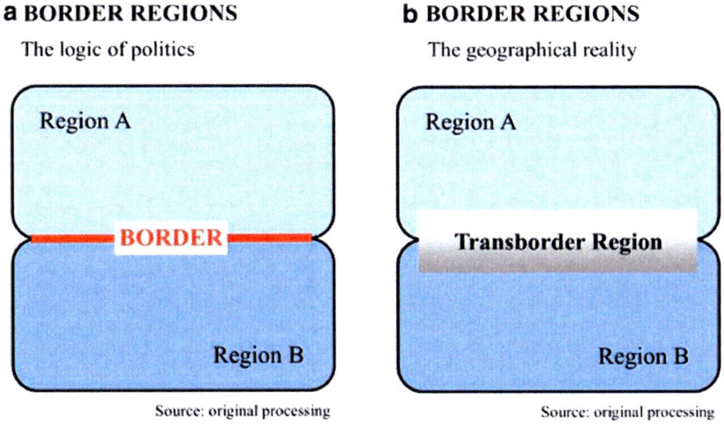

Fig. 7.1 Border regions. **a** The logic of politics. **b** The geographical reality. *Source* original elaboration

area, the cross-border region, which owes its marginality consolidated by being interlocked as a subset of the whole "union" of the two neighboring states.

The observation that borderlands tend to be marginal areas on both sides of the border, and that the condition of marginality tends to persist over time, is helpful to analyze the territorial problems of former borderlands, i.e. areas that had become part of a country's interior due to past wars and annexations. This is the case of one of the few landlocked regions of Italy, *Regione Umbria*, which was partitioned between the Byzantines and the Lombard Kingdom in the Middle Age, and later became a set of provinces of the Papal States bordering a variety of independent and semi-independent areas. As such, Umbria remained a borderland throughout the early modern and modern periods, until it became an internal area after the Italian unification. Throughout history, Umbria experienced great difficulties of aggregation that are present still today, shown in recurring political debates about possible partitions of its territory. Still today this region is a transition area, better yet—a border territory—between other Italian regions with stronger geopolitical, institutional, and sociocultural identities, namely Tuscany, Emilia-Romagna, and Lazio (Fig. 7.2).

The history of Umbria is a relevant case study to deepen our understanding of geographical phenomena. A borderland does not cease to be such simply because at some point the dynamics of history (and war) had pushed it in the interior of a larger polity, such as a state, an empire, or a federation of some sort. The dual character of geographic phenomena comes into play here, i.e. first, the influence of human action (i.e. the historical element) and second, the influence of nature (the physical-naturalistic element). The first character reminds us that, on the one hand, humans can modify certain features of territory relatively fast, but, on the other hand, they cannot alter relationships between the various features and between territories as fast.[3] Relationships—among institutions, social classes, settlements networks, and many more features of a given territory—tend to change very slowly, in a time frame that sometimes far exceeds the duration of human life. Such resiliency depends on the systemic nature of the various relationships established in a given geographical space, both by man and by nature.

As for nature, let us recall once again that boundaries are generally marked by morphology. This reconfirms the notion that marginal internal

[3] The notion of territory understood as bounded space organized by humans according to their logic and needs should be understood as a sort of large mosaic made up of many tiles in continuous movement. These tiles break down and recompose, welding together in different ways according to historical contingencies, which have the effect of opening or closing the fractures that separate the various geographical components.

Fig. 7.2 Italy, showing the regions and provinces discussed in the chapter. *Source* original elaboration

areas should be recognized through the identification of some discontinuity in the territorial fabric. It follows that perhaps we should speak not so much of border areas as particular cases of internal areas, but rather of internal areas as particular cases of border areas. In both cases, there is a problem of relations between the center and the periphery, which brings to mind the analysis carried out by Reynaud (1984). He developed the model of sociospatial classes to analyze the nexus between social inequality and place, highlighting the connections between inner-city neighborhoods and the notion of semi-periphery.

7.3 The Inevitable Reversibility of Geographical Conditions

It is widely believed that cross-border areas should find their strength in cross-border trade. Such trade should be a driving force for the economy, enabling capital to be accumulated locally over time. However, such broad generalization does not take into account the distinction between long-distance traffic, for which the border areas are at most a crossing point, and in respect of which local services develop in response to decisions taken outside the border region, and local traffic, which serves the two communities on either side of the border. Albeit important, such distinction is not always clear-cut, because there are situations in which local traffic serves a clientele coming from outside, such as, for example, the meat trade on the border between Italy and Austria. Farmers in the Austrian province of Tyrol sell meat to restaurants and supermarkets in the Italian South Tyrol, which ends up on the tables of tourists staying in the province of Bolzano (Valussi 1982).[4] Experience shows that only long-distance trade or at least trade serving communities outside the border region is truly developmental, in the sense that it promotes a substantial change in the economic fabric, attracting external resources which can trigger cumulative growth processes.

Local trade is based instead on local potential, both in terms of economic production and its absorption. The contribution to the local economy is limited to the supply of a limited range of goods at competitive prices. This does not mean, however, that in a regime of increasing openness, the possibility of adding even just the market potential expressed in the two countries cannot represent a politically rewarding objective for marginalized areas. However, it should be pointed out that these premises often give rise to illusions, insofar as adding two peripheral areas together does not in itself appear sufficient to give rise to a new centrality. Simply put, the majority of borderlands are marginal areas, suffering from resource extraction; therefore, it is unlikely that they can reverse the flows of resources just by trading with each other.

Returning to the initial observation, it should be noted that the evolutionary mechanism that leads from underdevelopment to development is part of a philosophical conception based on the idea of progress,[5] understood as "necessary" and potentially unlimited.[6] It is the myth on which the so-called modern world was founded, which, in the face of the failure of the project of real socialism in the last century and the ongoing crisis of the globalized private economy now reveals all its limits. Consequently, next to the progress it is necessary to consider also the category of regress (Forte 1965), which brings us back to the cyclical vision of G. B. Vico, based on historical courses and recourses. Translating Vico's philosophy in the perhaps more familiar language of analytical geometry, the function of any economic indicator used to measure changes in the territory should not look like a straight line, but rather like a curve that has several points of inflection over time.

All this reminds us that marginality and development, far from representing intrinsic conditions of the territories, are transient, a fact that only a crudely determinist vision of the relationship between man and territory prevents observers from seeing. It is worth remembering that in Italy this crude vision contributed to providing a scientific cover for discriminatory, even racist interpretations of the 'southern question'—i.e. the political debates concerning the underdevelopment of the South—that arose immediately after the national unification in 1861. These visions were imposed by the dominant political culture and could only be adequately challenged in recent times when the real economic situation of the various parts of the peninsula was measured and compared (Daniele and Malanima 2011).

[4] A cross-border trade agreement between Italy and Austria, called in Italian *Accordino*, regulated the "small border traffic" between the two parts of Tyrol before the two countries joined the EU, making it obsolete. Similar arrangements were the Udine Agreements, which applied to the border strip between Italy and Slovenia almost up to the entry of the latter state into the EU. (Battisti 1979, 1981, 1984, 1986). In both cases, the antiquated international trade mechanism called 'clearing' was used to mend at least in part the broken relations between the two parts of a historical region that was suddenly divided as a consequence of border shifts after the two world wars.

[5] The idea of progress, one of the theoretical assumptions of modernity, has moved away from its original religious formulation, proper to the Judeo-Christian tradition; it is used to explain transformative processes in different areas: ethical, naturalistic, cultural, sociological, political, and economic.

[6] According to an anti-capitalist interpretation, this idea is closely linked to the concept of capitalist development: "We are all so imbued with the self-justifying idea of progress created by this historical system that we find it difficult to even recognize the enormous negative historical features of the system" (Wallerstein 1995, pp. 41–42).

7.4 Borders as "Actors" of Regionalization

Let us now return to the border areas. In this case, the alternation of centrality and peripherality appears to be, in many cases, inherent to the dynamics of border shifts. The shifting of borders leads to changes in the positioning of territories and settlements relative to one another. This results in advantages for some areas and disadvantages for others. W. Christaller had already pointed this out in 1933 when he analyzed the central places along the new Franco-German border in the aftermath of the First World War (1980).

Broadly speaking, there are two main types of effects of border shifts on territories' dynamics of centrality vs. peripherality. First, if the displacement of the border is of limited entity in comparison with the size of the overall area involved, then there will be a fracture of the pre-existing functional regions. Second, if the change in the location of border has a greater magnitude, there will be a virtual translation of the entire region, which from a border area or cross-border area will be relocated within the larger geopolitical body. It will then become an internal area. Umbria experienced the second type of shift because it took the annexation of the whole Papal States by the Kingdom of Italy to first, transform its entire territory from a borderland into an internal area and second, to transform it into a constitutional region—i.e. one Italy's largest and most administratively relevant subdivisions.[7] From a theoretical point of view, such translations of geographic position represent a fundamental category of regionalization processes, parallel but opposite to the American frontier described by Turner (Battisti 2002). In this case, there is a topographical slide outwards of a set of human-made characteristics of spaces and places. Unlike Turner's, Umbria is a case of "successive occupation" of the territory, a phenomenon well known to North American literature, thanks to the work of biogeographers (Clemens and Shelford 1939) who, in the first half of the last century, had focused their attention on the notion of "ecological succession" (Park et al. 1925; Webb 1931).

Europe, during its multimillennial history, experienced a great variety of these and other typologies of regional shifts. Moving away from the example of Christaller's central places discussed above, analyses of crossroads cities show shifts in economic and political leadership occurred over the centuries among the port cities that flourished (and faded) along the entire coastal arc of the Adriatic Sea (Burghardt 1971). In this area, hegemony passed from Adria to Ravenna, to Venice, to Trieste, and, on the opposite shore, from Dubrovnik (Italian Ragusa) to Rijeka (Fiume) and Koper (Capodistria). From a geographical point of view, the process of growth and decline of crossroad cities is reminiscent of the development of seaside resorts development summarized by Miossec (1977) in his famous sequential model, which together with the model of Butler (1980) configures a sort of adaptation of Vernon's theory on the international life cycle of industrial products to the rise and decline of commercial cities. Beyond the superficial similarities between typologies of cities and typologies of industries, what matters most is the geopolitical transformations in the relations among territories induced by factors such as technological development and the related evolution of market forces.

In a nutshell, border shifts, alongside other factors such as the ones discussed above, have the potential to alter the relations between territories and between places in a given territory. One of the most salient European examples of territorial transformations due to border shifts comes from the area that stretches between the northeastern tip of the Adriatic Sea and the Eastern Alps, i.e. the former Austro-Hungarian province of the Littoral, or *Venezia Giulia* in Italian, which, after two world wars, found itself divided between Italy, Slovenia, and Croatia. The partition of the *Venezia Giulia* created new marginality by separating the countryside—assigned to the

[7] Italian geographers have repeatedly criticized the irrationality of the current subdivision of Italy into regions, resulting from the lack of attention to scientific arguments due to the urgencies of politics (Gambi 1963, 1977, 1995; Muscarà 1968; Treves 2004).

former Yugoslavia (today Slovenia and Croatia) —from its main urban areas, the cities of Trieste and Gorizia, which remained in Italy (Valussi 1972). Interpreted using Walter Christaller's central place theory (1933), the split between rural spaces and the corresponding urban areas caused a reduction in the level of centrality in the latter. So on the one hand, Yugoslavia built new cities on its side of the border (Nova Gorica, but in a sense also the new port settlement of Koper/Capodistria); on the other hand, the changed strategy of national investments has increased the centrality of inland cities such as Udine and Pordenone. Recomposing these fractures is hard, as shown in this book's previous chapter: in the grand scheme of things, the transformations experienced in Italy's Eastern border are not too different from those experienced in Umbria and in Bolzano, leading to the somber conclusion that breaking territorial units may lead to scars that last a really long time.

References

Battisti G (1979) Una regione per Trieste. Studio di geografia economia e politica. Università di Trieste, Udine, Del Bianco

Battisti G (1981) I movimenti "turistici" al confine italo-jugoslavo. Metodologie per la ricerca. Quaderni del C. S. E. Vanoni di Trieste, no 15, pp 1–13

Battisti G (1984) La collaborazione del Friuli-Venezia Giulia con la Jugoslavia e le prospettive di sviluppo. in: Atti 8° Convegno Scientifico Internazionale Alpe-Adria (Graz, 7-8 ottobre 1982): Interventi selezionati per la promozione dei rapporti economici nelle Regioni centrali della Comunità di lavoro Alpe-Adria. Incontro tra teoria e pratica, Facoltà di Economia e Commercio, Università di Trieste, pp 107–223

Battisti G (1986) Le relazioni frontaliere sul confine tra Friuli-Venezia Giulia e Slovenia 1982–1985. Tra crisi ed evoluzione. In: Atti 10° Convegno Scientifico Internazionale Alpe-Adria (Trieste, 24–25 ottobre 1986). Facoltà di Economia e Commercio, Università di Trieste, pp 61–103

Battisti G (2002) Tra confine e frontiera: la regione "mobile". In: Battisti G (eds) Un pianeta diviso. Contributi alla geografia dei popoli e dei confini. Dipartimento di Scienze Geografiche e Storiche dell'Università, Trieste, pp 101–114

Burghardt AF (1971) A hypothesis about gateway cities. Annals, A.A.G., pp 269–285

Butler RW (1980) The concept of the tourist area life-cycle of evolution: implications for management of resources. Can Geogr 24(1):5–12

Cencini C, Dematteis G, Menegatti B (eds) (1983) L'Italia emergente. Indagine geo-demografica sullo sviluppo periferico. Franco Angeli, Milano

Christaller W (1980) Le località centrali nella Germania meridionale. In: Pagnini MP, Milano FA (eds)

Clemens FE, Shelford VE (1939) Bio-ecology. John Wiley & Sons

Daniele V, Malanima P (2011) Il divario Nord-Sud Italia 1861–2011. Rubbettino, Soveria Mannelli

Forte F (1965) Un caso di economia in regresso relativo: l'economia triestina. Rivista Internazionale Di Scienze Economiche e Commerciali 12(8):791–814

Gambi L (1963) L'equivoco tra compartimenti statistici e regioni costituzionali, Faenza, F.lli Lega

Gambi L (1977) Le "Regioni" italiane come problema storico. Quaderni Storici, 34, pp 275–298

Gambi L (1995) L'irrazionale continuità del disegno geografico delle unità politico-amministrative. In: Gambi L, Merloni F (eds) Amministrazione pubblica e territorio in Italia. Il Mulino, Bologna, pp 23–34

Leone U (ed) (1986) La rivalorizzazione territoriale in Italia. Franco Angeli, Milano

Leone U (ed) (1988) Valorizzazione e sviluppo territoriale in Italia. Franco Angeli, Milano

Miossec JM (1977) Un model de l'espace touristique. Espace géographique, 6, pp 1–48

Muscarà C (1968) Una regione per il programma. Marsilio, Venezia

Park RE, Burgess EW, McKenzie RD (1925) The City. The University of Chicago Press, Chicago

Reynaud A (1984) Disuguaglianze regionali e giustizia socio-spaziale. Unicopli, Milano

Rostow W (1960) The stages of economic growth: a non-communist Manifesto. Cambridge University Press, New York

Sommella R (2017) Una strategia per le aree interne italiane. Geotema 21(55):76–79

Treves A (2004) I confini non pensati. Un aspetto della questione regionale in Italia. Annali della Facoltà di Lettere e Filosofia. Università degli Studi, Milano, 57, 2, pp 244–264

Valussi G (1972) Il confine nordorientale d'Italia. Trieste, Lint

Valussi G (1982) La collaborazione economica del Friuli-Venezia Giulia con l'Austria e le prospettive di sviluppo. Istituto di Geografia, Udine

Wallerstein I (1995) Historical capitalism with capitalist civilization, London and New York, Verso; Italian edition (2012) Capitalismo storico e Civiltà capitalistica, Trieste, Asterios

Webb WP (1931) The great plains. A study in institutions and environment. Ginn and Company, Boston, Mass.

Part IV

From Trieste to the World: Deploying Triestino Geographical Thought to Grand Geopolitics in Europe and Beyond

8 The Reshaping of German-Yugoslav Space from a Middle European Point of View. Paper Presented at the 2nd IBRU Conference, Held in Durham, UK, on July 18–21, 1991

Abstract

This chapter opens a new section of the book, dedicated to the application of Battisti's realist, empirically driven historical geographies to larger scope questions of Europe and global geopolitics. It consists of the notes of a conference presentation delivered in England in 1991. The chapter reveals the author's real-time considerations and predictions about the transformation of Europe following the end of the Cold War. It showcases the difficult interactions between the unspoken norms of international geography and Battisti's Italian tradition—at the time it was rejected from publication. Reading it as a methodological piece, it shows some of the blind spots in international geography and the benefits of keeping an open mind to 'other' traditions. Reading as a historical document, it shows the volatility of European geopolitics during the months leading to the collapse of the USSR and Yugoslavia. These include the possibility of Italian involvement in the former Yugoslavia, the explosion of conflicts between ethnic groups way beyond the wars in Yugoslavia of the 1990s, and the possible regional integration projects alternative to the enlargement of the European Union.

Keywords

Cold war · USSR · Yugoslavia · Germany · European Union · Realism · Geopolitics

8.1 Introduction: Reading Thirty Years Old Notes

Looking for materials to base this book, I found this conference paper I delivered at the second I. B.R.U Conference in Durham, UK, in July 1991. Reading it almost exactly thirty years later, Christian and I felt this is the right place to introduce my take on European and global geopolitics. First, the paper shows the extent to which 'my' geography is the product of a tradition that has taken a different path of development from Christian's Anglo-American colleagues. Indeed, this paper was a clash between these two traditions. At the time, 'cultural turn' Anglo-American human geography was not sympathetic of the realist and historicist approach expressed here, which included the insistence of viewing states as organic, reified entities, rather than viewing them in the prism of 'meaning,' or as bundles of relationships between structure and agency (Jessop 2001).

C. Sellar and G. Battisti, *Geopolitical Perspectives from the Italian Border*, Historical Geography and Geosciences, https://doi.org/10.1007/978-3-031-26044-5_8

Needless to say, the colleagues did not like the paper, so I did not pursue publication.

Second, British colleagues' dislike notwithstanding, this paper's 'old school' geopolitics could analyze transformations in the system binding European states in almost real time. It was written right before the final collapse of the bipolar world that had determined the fate of Europe since WW2. I delivered the paper on July 19; German reunification was less than one year old; a few days earlier, at the end of June, Slovenia and Croatia had declared independence from Yugoslavia. On July 7 both countries signed the Brioni Agreement with Yugoslavia to postpone independence for three months (Debié 1997); in August a coup in Moscow marked the beginning of the final collapse of the Soviet Union; in October war erupted between Croatia and the rest of Yugoslavia; in December the Soviet Union ceased to exist. Amid such upheavals, more closely reminiscent of the realpolitik of the nineteenth century than the ideological Cold War of the 20th, the realist approach to the discipline made sense, and the paper offered considerations about the risks of ethnic conflicts and nationalisms, the re-emergence of Germany as the powerhouse of Europe, the stabilizing role of the European Union, and future political tensions that partly reflected the events of the following two decades. Some of the predictions reported here did not materialize—such as the risk of conflicts over Kaliningrad in the Baltic Sea, or the unification of Kosovo and Albania—but they nevertheless show how volatile the European situation was in 1991. Indeed, the unfortunate events of the new millennium show that volatility has persisted. In retrospective, long-term instability in Europe was to be expected. From a geopolitical perspective, geographical positions retain their specific characteristics for very long periods, until the overall system of power relations undergoes substantial changes. The "hot spots" therefore remain such and can flare up at any time. If currently the presence of a large US Army base seems to represent a factor of stability in the increasingly critical situation in Kosovo, the recent decision by Lithuania (June 2022) to stop the transfers of goods between Russian territory and its exclave of Kaliningrad presents again, albeit from a different perspective, the concerns I raised in 1991. Moreover, some of my concerns were publicly validated, albeit a few years after my presentation. For example, Italian analysts confirmed the secret negotiations between Italian and Yugoslav leaders in the two countries concerning the return of territories Italy lost after WW2 in exchange for support against Croatia, risking an Italian involvement in the upcoming Balkan wars (Fubini 1996).

Because of both the timeliness of the analysis and the nature of this paper as an encounter between geographical traditions, my co-author and I deemed it worth publishing it unedited for either language or content. The only changes consist in explanations, in brackets, of acronyms or situations that were common knowledge in 1991, but may not be so obvious to international readers in the new millennium. We decided that leaving these notes as close to the original as possible is a powerful representation of the challenge to interact in a second language with a different scientific tradition, a specific example of writing geopolitics from Trieste, and a firsthand testimony of the year that transformed Europe.

8.2 Paper Presented to the 2nd I.B.R.U. Conference (Durham, 18–21 July 1991). Foreword

At the origin, the title of this presentation was largely tentative. My former intention was to speak either of the German case or of the Yugoslav one. The events which have followed stimulated a wider reflection on both sides of the problem. They are strongly integrated in the perspective of the growth of a vast community of people ethnically homogeneous, with a periphery occupied by a constellation of populations highly unable to reach a reasonable agreement. This is true for the Baltic as well as in the Balkans.

I would like to draw your attention to the similarity between the process on the road and some past events which profoundly influenced the European political map. I will try to synthesize the process in being.

A brief introduction is needed. I live and work in Trieste, a city whose fortunes were marked by the evolution of the borders (Battisti 1986). I am a testimony of these processes, having experienced the military occupation and the partition of the region, i.e. something like in the postwar Berlin. After the Treaty of Osimo (1975) I spent much time to study the border economy and its bases, which resulted in more than 500 pages printed on the matter (see Bibliography). Maybe the results could be of more interest, but having been published (for the greater part in Italy), I found it not worth repeating them. I am at your disposal for further information.

Now, let me make my assumption. States are not immovable entities, they are born, transformed, put to death. Their fate depends upon the recognition they may obtain from the community of other states. In the economy of international affairs, they shall be entitled to this recognition, and deserve it, as long as they are perceived as useful. Every state has certain specific functions to perform, which are largely related to their geographic configuration. The birth of a new border is generally tuned to these considerations.

8.3 The Yugoslav Question

I now wish to illustrate my thesis. Yugoslavia was established after WWII as a buffer state between Eastern and Western Europe. Nowadays, the confrontation between democracy and communism has come to an end, and consequently the Yugoslav state has lost its importance. Nothing serious, you may say. But what is serious indeed is the fact that this state completely lacks of homogeneity in respect of what are usually considered the vital elements for a state: i.e. ethnicity, culture, and economy. No ethnic group possesses either the majority of the population or is significantly settled over the whole country. We must argue that for Yugoslavia the existence of forces pushing against it from opposite directions was vital to its internal cohesion. In lack of pushes, the centrifugal forces regain momentum. As usually, all of that is taking place in a time of widespread economic crisis. In this paper, I shall not deal with the economic reasons and implications, which would require a completely different approach. I shall concentrate therefore only on the geopolitical side of the question.

Let us now give a look at the future: the point in question is "Can Yugoslavia just dissolve?" and if the case, "How can this happen?" At first glance, Slovenia could easily get out of the game. It possesses a clear-cue ethnic border with Croatia and an enviable ethnic homogeneity (90% of the population is Slovene). By paradox, Serbian leaders could profit from its secession, which would leave Croatia no longer backed in its quarrel with the republics of the South. Unfortunately, no easy solution may be envisaged for the Croatian-Serbian confrontation, because of the intertwining of populations in the whole area. Centered in Bosnia-Herzegovina, it extends from the former Austrian military border[1] to the N-W [northwest] part of Serbia. In this case, a partition of Yugoslavia would lead to a long-standing political struggle between the two main opponents, i.e. Serbia and Croatia. Both of them would be engaged in a complicated gerrymandering in order to confine the other ethnicities to the role of harmless minority. This perspective is unacceptable to both, the last issue being the resort to arms.[2] Maybe a final exchange of population could take place, at a scale (and with the costs) largely experienced in the Greek-Turkish struggle after WWI.

Things being what they are, it would be of little use for the Serbians to let the Slovenes go away. Once established the principle of "free secession," nobody could stop the Albanian

[1] On the origin of the mixed population area, see Blanc (1957) and most of the works recently appeared in the "Geographical Papers", University of Zagreb, 8, under the heading Geopolitical and demographical issues of Croatia. It gives a thorough information to the whole problem of minorities in Croatia, and to the scientific questions involved.

[2] The events are occurring at a growing pace, therefore parts of this paper may become obsolete even at the date of its publication.

community living in Kosovo from joining the Republic of Albania. A very underdeveloped country, we know; but the Albanians are not discriminated there, and the regime is turning for the better.

This secession will certainly take place, sooner or later, in my opinion. It is in the logic of the numbers. So the Balkans (not only Serbia) shall face a new state of more than 5 million people (more than Croatia). But that day the whole political balance of the peninsula shall be altered.

I suppose that everyone will agree with me (and Gilas 1991) in considering extremely hazardous helping changements of such magnitude and scale in the Balkans. Once Serbia has been weakened, it is dubious that Hungary will refrain from requesting its part of Vojvodina (about 600,000 inhabitants), as well as the Baranja from Croatia. Then, the attention could be turned toward Transylvania. As for Bulgaria, its official policy looks to the Yugoslav republic of Macedonia as an integral part of its territory. The scenario reminds the Balkan wars again (Fig. 8.1).

8.4 Toward a New Danubian Confederation?

The mentioning here of Hungary recalls the attitude of Austria toward the Slovenian and Croatian issues: they are too much enthusiasts, in Vienna, to give the secessionists political recognition, forgetful of the troubles with the Slovene minority in Carinthia and Styria. The Yugoslavs repeatedly[3] requested 2470 km^2 [square kilometers] of the first, and 130 km^2 of the latter. An unofficial census was made of this people, who like those in Italy considered it a repressive measure. There is some evidence that in Vienna they favor the disintegration of Yugoslavia. Only then, they will speak of minorities again. Slovenia and Croatia were once part of Austria, at the very end.

This behavior reminds the *Sonderbund* War in 1847. However, in that case, the Austrian policy failed its scope, and the country involved—Switzerland—succeeded in establishing strong centralized institutions. The Austrians seem eager to repeat the success of [Western Germany's] President Kohl in absorbing the G.D.R. [East Germany] just overnight. It is a happy time for Austria. The Soviet retreat from Eastern Europe canceled the principal opposition to its integration into the EEC [European Economic Community, today's European Union]. Economic ties are being strengthened with Hungary, a country where some people want to join NATO (and the same is happening in Czechoslovakia). The small alpine republic [Austria] catches a glimpse of a new confederation along the lower-medium Danube basin, inclusive of Slovenia and Croatia.

It is an illusion, of course. Behind Austria, there is—more or less openly—the new German state (some 78 million people, the largest political entity in all of Europe). Austria is an economic satellite, with 60% of its imports and 40% of its exports depending on the neighboring country. However enlarged, the Austrian political space would be a dependent one. Since its origin (1978), in the Alpe-Adria Working Community, the Bavarian *Ostpolitik* has been meeting Slovenian and Croatian *Westpolitik* (see Fig. 3.1).

8.5 New Problems for Old Minorities

After the secession, these small states should face big economic problems, which will force their leaders to rely on nationalism again.[4] Number is strength also in the economic competition. Croats are worried about the long-standing loss of population (Nejasmic 1991) because of emigration and mixed marriages (Sterc 1991). All these losses are felt as a national disaster.

With a population of less than 2 million people, Slovenia ought to call each one of its sons to

[3] Notwithstanding the plebiscite of Klagenfurt in 1920 (Dami 1976: 98–99, 264–265).

[4] The growth of nationalisms in Yugoslavia is well documented by Poulton (1991).

Fig. 8.1 Geopolitics in progress: the Yugoslav drama. *Source* original elaboration

cooperate with the mother country. The small minorities in Italy and Austria (less than 100,000 people: Valussi 1974) would inevitably be involved. This could fuel frictions with the two neighboring countries, at the same time destroying the very sources of their prosperity. In Italy, the Slovene community has been allowed to establish some banks, and to gain the virtual monopoly of border commerce (Battisti 1984, 1986), worth some 250 million Lg. [pound sterlings] per year (see Fig. 3.2). In the case of an independent Slovenia, all these activities would migrate toward the new frontiers. Moreover the Italian border region *Friuli- Venezia Giulia* would be largely deprived of the investments made during the last decade by Slovenes and Croats (estimated value: more than 2 billion Lg.). There were rumors in the past, that even Marshal Tito's wife owned some shops in Trieste, either as an active or silent partner. The flight of these capitals (mostly illegal) is part of the quarrel with the Serbs.

On the other side, the Italian minority in Yugoslavia would be divided by a new border, following the ethnic line separating Slovenes and Croats in Istria: a very recent border, created in 1947 along the River Dragogna [this is the current international border between Croatia and Slovenia in the Istrian peninsula]. In the last decades, the treatment reserved to this minority (15,000 units in 1981, but see Note 9) has been changing. In Slovenia, an Italian has recently been appointed Mayor of Capodistria (Koper). In the Croatian part of Istria and in the Fiume (Rijeka) area on the contrary, where the great majority of this community is still living, it is impossible to receive the broadcasts from Telecapodistria [the only Yugoslav television network broadcasting in Italian]. This is a subsidiary of Ljubljana Television and can be seen in the West as far as Sicily. In the future, there would be no place for Italians in a Croatian state born out of a nationalistic revival.

Until now, the Italian minority survived in a sort of unwritten exchange of the Slovene one in Italy, where there are no Croats as yet (not to count some historical ethnic "islands" in the

South). Inside the country, they could appeal to the internationalist element contained in the socialist ideology professed by Tito's political building. But what role could they play in a capitalistic, right-wing state? There are plans to resettle in Istria 20,000 Croats from Romania (Poulton 1991, 97). To avoid such a future, some people rely their hopes in an autonomous region, which should be trilingual, and guarantee a double nationality status at least for the Italians that live there. A region open to foreign investments, tourism, and shipping. This could prove to be an original solution, like the transformation of Bosnia in Cantons, a proposal that nobody seems to accept at present.

However, one point above others must be clearly stated. The life of all minorities is endangered by the present turmoil and havoc. In Italy, the neofascist party claims that the Osimo Treaty has decayed after the declarations of independence by Slovenia and Croatia. It was indeed signed with the Federation, a country which is likely to disappear.[5] We all perfectly know (also thanks to Prof. Tyranowsky[6] that such a thesis is a nonsense for the international law. Notwithstanding that, the Serbian leaders (B. Mihasic, Ceikovic) contacted Secretary G. F. Fini [the leader of the Italian right-wing party] posing the question of the Italian rights in Croatia.[7] This is a classic move, a poisoned letter sent not directly to the secessionists, which threatened an allegiance of minorities susceptible to give fire to another region. On the other hand, the Slovenian media spoke of popular resentment against Italy, for not backing Ljubljana.[8]

From these games, the Italians—and not only those living in Yugoslavia—can only lose. No more blood, please; no more refugees. Actually, also Croatian scholars (Nejasmic 1991, 72–2; Zerjavic 1991, 94) now recognize that at least 195–230,000 Italians were forced to leave their homes at the end of the war, their lives being in danger.[9] No more bloodshedding, please. Don't involve us once more in your own troubles.

8.6 Is the German Question Really Ended?

Other peoples can afford trying to reverse history, and to pay the price for it. The disintegration of the socialist order in Eastern Europe has caused a massive movement of technicians, bankers, businessmen, now beating the tracks open since the Middle Ages by knights, monks, farmers, craftsmen, to colonize a vast, underpopulated territory (Turnock 1989). A colonization that most of the time was peaceful.

[5] The party's position is the following: Istria was given to Yugoslavia, one of the winning states of the WWII. On the contrary, Croatia fought on the German/Italian side. To consent Croatia to keep Istria any longer would be like having given South Tyrol to Austria in 1945. The fact that Croatia lost the war and was substantially subjugated by Tito's forces is well documented in Novak (1970), and is revealed by the present Croatian attitude to the Serbs. The Liberal Party (which is a partner in the ruling coalition in Rome), recently approved, during its national convention, a document asking for a renegotiation of the Osimo Treaty. Both parties, however, are facing a serious loss of consensus.

[6] Paper presented at the same conference.

[7] See Il Piccolo (Trieste daily paper), July 14, 1991.

[8] V. Verboseck, one of the mayors of Ljubljana, would have stated: "If we gain independence -and we'll get it - we can put everything into discussion, also the lands that once were Italian" (Il Piccolo, July 18, 1991). These statements are highly dubious, but no wonder whether, both Serbs and Slovenes, in their search for support, would offer Croatian Istria to Italy.

[9] Their figures match very well with estimates by Schiffred (1990, 259). The questionable criteria used by the majority groups against the Italians is evidenced by the census data. From 1948 to 1961 (when the movement has virtually ended) the Italian population in Croatia dropped from 76,093 to 21,103, that is, 55,000 people less. Now Nejasmic (1991, 73) states that after this year (1948) about 110,000 people moved to Italy, twice the official data! The author correctly points out that not only the Italians have left the country (as is often maintained), but also a large number of Croats. This was to concentrate in the illegal emigration, which is most intensive in the period 1955–1970. The corresponding intercensal drop being of 12,213 people, even counting all of them as Croats, it is evident that Yugoslav census underestimated the Italians by some 50%. On this basis, it is rather difficult, for an independent scholar, to take for granted whatever claim coming from the opposing nationalists in Yugoslavia. Enlightening is also the paper by Sterc (1991).

At the opposite side of Europe, other peoples are now getting excited. The Baltic populations hope to regain independence. These ill-advised uprisings may cause unwanted consequences too (see Fig. 3.3). A look at the map allows to note an exclave—the Russian autonomous territory of Kaliningrad—laying between Poland and Lithuania. The urban area has a population of 800,000, and with its military bases it plays a strategic role in Moscow's security system. The historical name of this city, we all know, is *Koenigsberg,* [a German name for a former German city, where still remains the grave of Immanuel Kant, which implies the possibility of Germany reclaiming the city] and its land forms part of Eastern Prussia [the cradle of the German empire until WW2]. There has been a lot of talk recently of turning it into a free port, an international city still belonging to the Soviet Union but in some way united with Germany (Nava 1991, 103). No definite agreements have been signed yet, but this Baltic Hong Kong might give the area a new chance.

The introduction of capitalistic production systems in the USSR will need some growth poles. Nothing better than an old port city. And as for the name, even Leningrad citizens wish to return to a past they never experienced [less than two months after delivering this paper, in September 1991, Leningrad re-acquired the old, German sounding name of St. Petersburg through a citywide referendum].

8.7 Will the Soviet Union Disintegrate?

Lee us now figure our such developments in light of the secession of the Baltic states. This would probably cause, as in 1918, the generalized collapse of the Russian state. Considering these assumptions and implications what would ultimately be like the final status of Kaliningrad? Shall it be disputed by Poles and Lithuanians, or is it more viable for the birth of a very new entity, more and more linked to Germany? If this happens, what about Klaipeda (Memel) and the rest of former German lands that Poland was forced to incorporate in 1945? (Maiklejohn 1983). Is there anyone who thinks that a highly indebted country [Poland] can permanently extend over a surface of 312,688 km^2. Besides the biggest nation in Europe [Germany], where on a slightly larger surface (357,020 km^2) is living more than twice its population?

The *Koenigsberg* case, in the perspective of a Soviet garrison called to defend a state no longer existing, could produce a new territorial discontinuity in the Baltic. A spatial configuration this, which proved many times to be explosive (Dami 1976, 114, 147–151, 153, 159–162). As for the real interests of the communities involved, don't forget that in 1982 the GDR [East Germany] resorted to open a rail ferry line from Mukran (Isle of Ruegen) to Klaipeda because of the behavior of the Poles (Scharmann 1990, 199). If I were a Pole, soon after the Lithuanians become independent I would hurry to buy some bags.

Violence is not necessarily to be associated with these developments. Lands are generally traded. German people are currently buying real estate (mainly villas on the lakes and seaside resorts) in Northern and Central Italy. It is interesting to note, moreover, that the "great divide" between Northern and Southern Italy, in economic terms, is a replica of the Southern border of German investments before the First World War (Hertner 1984, 275–331).

In the "Neighbourhood and Cooperation Treaty" signed last June, the Polish government agreed to recognize and protect the German minority still living in the country. Having obtained a free entry in Germany for Poles, how could they prevent German natives to come back to Silesia, Pomerania, Eastern Prussia [former German lands ceased to Poland after WW2], the day Poland will have joined the European Community?

As for the core of the German space, the reunification was a gamble, and it took place at the first available opportunity. It could have been the last one, too. The sense of nationality is a cultural dimension of man, and it requires people who have received the right cultural inputs during their formative years. It is enlightening to hear the position of the younger generations. Says Matthias Bauman, a student of Geography at Bonn

Fig. 8.2 Patterns of uncertainty on the Baltic shores. *Source* original elaboration

University[10] "For me, the East is like Austria. There they speak German too. But I feel no tie. I have no sense that Eastern Germans are my countrymen. Older people feel this connection more, because they lived through the division. I grew up with the fact that GDR was another country. I think this is typical for my generation. No matter, the game—this part of the game—is over." (Fig. 8.2).

[10] Quoted from Time International, 1991, 26:13.

8.8 The Role of the European Community

From North to South, the periphery of the old German space (defined as the area of once scattered German settlement) is being fragmented into smaller cores and even smaller peripheries. A disintegration which Italian geographers place within the framework of deregionalization (Turco 1984). In light of the whole process, this is the step immediately prior to its reversal, when centripetal forces regain control on the system. The very problem is, which kind of control, which kind of political forces will operate?

The crucial factor is time. A satisfactory reorganization can't take place in a hurry, especially in a period when the world economy is facing severe problems. In Italy, we say, "Hurry is a bad mistress for politicians". The temptation to use nationalism to capitalize peoples' consensus is always open, as the Yugoslav experience is teaching us.

The next steps of the European Community could offer a safe landing from the turmoil. The enlarged EEC could act as a catalytic factor, the constructive phase in this dramatic reshaping of the continental political map. In general terms, the challenge is, "will the whole EC succeed in gaining enough control of the Eastern countries, as to avoid a massive economic and political colonization by Germany?" The "steady state" would therefore be, a hierarchical system, where each partner country would be integrated at its proper level.

I am conscious that this presentation could sound as an exercise in geopolitics. Policies being the art of doing things at the very moment they are possible, the future is quite unpredictable. In any case, I presented a theoretical analysis, not my personal wishes.

References

Battisti G (1984) La collaborazione economica del Friuli-Venezia Giulia con la Jugoslavia e le prospettive di sviluppo [The economic collaboration between Friuli Venezia Giulia and Jugoslavia, and the perspectives of development]. VIII Convegno Scientifico Internazionale Alpe-Adria, Graz October 7–8, 1982). Università degli Studi, Facoltà di Economia e Commercio, Trieste, pp 107–222

Battisti G (1986) Le relazioni frontaliere sul confine tra Friuli-Venezia Giulia e Slovenia 1982–1985 [Cross-border relations between *Friuli-Venezia Giulia* and Slovenia 1982–1985]. Decimo Convegno Scientifico Internazionale AlpeAdria. Università degli Studi, Facoltà di Economia e Commercio, Trieste, pp 61–103

Blanc A (1957) La Croatie occidentale [Western Croatia]. Institut d'Etudes Slaves de l'Universite, Paris

Dami A (1976) Les frontieres europeennes de 1900 a 1975 [European borders from 1900 to 1975]. Edition Medecine et Hygiene, Geneve

Debié F (1997) Paper tiger: Europe and the crisis in the former Yugoslavia (1991–1996). Camb Rev Int Aff 10 (2):122–140

Fubini F (1996) Quando Fini sognava Istria e Dalmazia [When Mr. Fini dreamed of Istria and Dalmatia] Limes. Rivista Italiana di Geopolitica. https://www.limesonline.com/cartaceo/quando-fini-sognava-istria-e-dalmazia?prv=true. Access 22 Oct 2021

Gilas H (1991) Ecco perche la Federazione deve restare unita [Here is the reason why the (Yugoslav) Federation must stay united]. Corriere della Sera, 12 July (translated from: New Perspective Quarterly)

Hertner P (1984) Il capitale tedesco in Italia dall'unita alla prima guerra mondiale [German capital in Italy from the reunification to the First World War]. Il Mulino, Bologna

Jessop B (2001) Institutional re (turns) and the strategic–relational approach. Environ Plan A 33(7):1213–1235

Maiklejohn TS (1983) Poland's place in Europe. Princeton University Press, Princeton

Nava H (1991) Rostock and the eastward ports. Abitare 298:100–105

Nejasmic I (1991) Emigration from Croatia—a quantitative approach to a century long process. In: Crkvenčić I (ed) Geopolitical and demographical issues of Croatia, Zagreb. Dept. of Geography, University of Zagreb, 61–82

Novak B (1970) Trieste, 1941–1945. The ethnic, political and ideological struggle. The University of Chicago Press, Chicago

Poulton I (1991) Balkans. Minorities and states in conflict. Minority Rights Publications, London

Scharmann L (1990) Further remarks on the urgent modernization of the Central European transport networks. Some special cases concerning Eastern Germany. Geojournal 22(2):195–203

Schiffred C (1990) La questione etnica ai confini orientali d'Italia [The ethnic question at Italy's Eastern border]. In: Verani F (ed) Edizioni Italo Svevo, Trieste

Sterc S (1991) Ethnic origin of 'Yugoslavs' in Croatia". In: Crkvenčić I (ed) Geopolitical and demographical issues of Croatia. Dept. of Geography, University of Zagreb, Zagreb, pp 143–166

Turco A (1984) Regione e regionalizzazione [Region and regionalization]. Franco Angeli, Milano

Turnock D (1989) Eastern Europe. A historical geography 1915–1945. Routledge, London

Valussi G (1974) Gli sloveni in Italia [The Slovenes in Italy]. Lint, Trieste

Zerjavic V (1991) The losses of Yugoslav population in the second world war. In: Crkvenčić I (ed) Geopolitical and demographical issues of Croatia. Dept. of Geography, University of Zagreb, Zagreb, pp 83–107

9 Europe. The Many Reasons of an Epoch-Spanning Crisis. A Long-Term Geohistorical and Geoeconomics Analysis of the Obstacles to European Integration

Abstract

This chapter discusses the structural weaknesses of the European Union as a project of continent-wide political integration. In so doing, it performs a long-term analysis of the process of State building in Europe to argue that the multifaceted legacy of the mosaic of polities that have constituted Europe over the centuries hardly fits the rationalistic administration typical of the modern state, and even less the attempt to root supernational unity on the same rationalistic principles. This chapter's topic is relevant to discussions of the territorial trap (Agnew in Rev Int Polit Econ 1:53–80, 1994) the historical development of territorial states, and the relationship between geopolitics and the economy. This is also the most overtly political piece of this book, taking positions on the nature and structure of modern states that are close to the classics of conservative political thought.

Keywords

European Union · Geopolitics · Geo-economics · Territorial trap · Conservatism · Historical geography

9.1 Introduction: A Long-Term Analysis of European Geopolitics

This chapter continues the reflections on Europe's geopolitics, deploying the methodological tools discussed in the first section of this book, namely careful attention to historical geography, the preservation, whenever possible, of an organic vision of geography as the convergence of its various subfields, which results in keen attention to the mutual influences between history, realist geopolitics, and geoeconomics. The outcome is a long-term analysis of the evolution of European states, a process fraught with contradictions that reflects in the structural weaknesses and the crisis of the European Union in the early twenty-first century, which this chapter interprets as the latest chapter in the evolution of European polities. In so doing, this chapter considers issues discussed by a variety of Anglo-American scholars, including, but not limited to, the peculiar position of Europe in the imagined subdivision of the surface of Earth in continents (Lewis and Wigen 1997), the fleeting nature of territorial states (Agnew 1994, 2017), the historical processes leading to their establishment as a hegemonic form of politi (Biggs 1999), and the relationship between geopolitics and geoeconomics (Moisio 2018).

C. Sellar and G. Battisti, *Geopolitical Perspectives from the Italian Border*, Historical Geography and Geosciences, https://doi.org/10.1007/978-3-031-26044-5_9

However, the paper also takes positions rarely heard in geography, closer to the classics of conservative political thought: the comments on the administrative reforms in revolutionary France are reminiscent of Edmund Burke's condemnation of the destruction of the variety of human experiences brought by these reforms (Burke 2009, pp. 149–150). The observations on the authoritarianism of modern nation-states remind of Irving Babbitt's critique of majority voting as an oppressive form of politics (Babbitt 1924, pp. 70–96); and the importance of religion as a crucial part of public life echoes the first of Russell Kirk's six canons of conservative thought—i.e. the 'belief in a transcendent order… political problems, at the bottom, are religious and moral problems' (Kirk 2001, p. 8). The outcome is unusual, perhaps politically incorrect, but with elements of originality that may attract some readers' attention.

9.2 Between History and Geography

Paraphrasing what Chancellor Metternich said about pre-unitary Italy, Europe remains a geographical expression. For the geographer, it is a (small) peninsula of Asia, although as the scale of the map increases, we can appreciate how it consists of a plurality of peninsulas, each with its distinct individuality. As with cartography, distance plays a fundamental role in the anthropic character of the whole: things are always better understood from an external perspective. The myth of Plato's Cave proposes a principle of universal validity. From a distance, details blur, giving way to a unitary image that seems to suggest the existence of a homogeneity that is the result of an optical illusion, like the "channels" that in 1877 Schiaparelli believed he had discovered on Mars.

Historically, culturally, and economically, Europe is much more than a peninsula. Its influence—which is then the set of various influences, coming from a system all in all unitary in which the circulation of ideas, men, and goods has always been uninterrupted—has indelibly marked the face of the planet. Etymologically, it means "west:" one of the possible etymologies derives from the Semitic word *ereb*, which meant everything West of Syria. Later, from the viewpoint of Greek navigators that followed the regular path of the sun, the term acquired the meaning of the end of a trajectory, the horizon on which Greek civilization moved, gradually expanding toward the mysterious "Sea Ocean," i.e. the Atlantic. Conceptually, its destiny is explained by its geographical location: it is the boundary-place reached by the Mediterranean peoples. This is a legacy so strong that it lasted until the end of the Cold War, in which the West and the East once again became two conflicting parties. Exactly as it began 25 centuries ago with the struggle of the Greek city-states against the Asian empire of the Persian kings (Chabod 1977).

From the classical, medieval, and even Renaissance, point of view the East was configured as the place of birth: God placed the first men in a garden "in the East" (Genesis 2:8). Even leaving aside the migrations of peoples, which alternate origin and destination over the millennia, it is no coincidence that the religions that profoundly influenced the formation of Europe as a distinct sociocultural space—Christianity, Judaism, and Islam—came from the East. This is a fact that the drafters of the European Constitution (presumptuously approved in 2004 by the governing elites, but rejected by popular vote in the following years) decided to censor, excluding from this document the Judeo-Christian roots of Europe. For these alleged "wise men", Europe—of which no definition is given—is a continent of immigrants. Essentially devoid of ideas and culture of their own, they would have come from everywhere to a place where, almost magically, the ideology of Enlightenment emerged, which would represent the key element of "Europeanism" (Battisti 2011). Today we can understand how it contained a rather explicit invitation to open the doors to wild immigration. The community design was reduced to a simple "blender of people", a concept found in the Preamble of the draft approved in 2003, which was almost totally expunged in the text approved by the heads of government the following year. The very presence of such a concept during high-level negotiations suggests a strategy of population replacement.

In the current contingencies, this is a homemade version of the legendary "melting pot" from which the mythical *homo americanus*[1] was once believed to spring. According to these intellectuals, committed to discovering new "human rights", the construction of Europe should then pass through the cancelation of the only principles and values that all European peoples share, consciously or not. Instead, it should not be forgotten that Christianity had been a crucial ideal, as well as a symbol, accompanying European endeavors up to the Enlightenment period. The cross was the symbol of the balance achieved in the Middle Ages between the throne and the altar. During the colonial period, it was not only an ideological justification for conquest but above all a testimony to the trust that men of the sea, accustomed to risking their lives at all times, placed in God.

The importance of Christianity was such that for most of the medieval period what today we identify as Europe was most commonly known as 'Christendom.' Today, it seems that this awareness has disappeared, thanks to the atheistic ideologies elaborated in Europe since the eighteenth century and which came to power in the twentieth century. Unfortunately, the same estrangement from the faith of the fathers also seems to be taking place on non-European continents where European emigration has taken over from the native populations.

[1] The concept of *homo americanus* was launched by Israel Zangwill in a drama first performed in 1908, in which he describes the loss of identity of Jews who had arrived in the "new world" (Zangwill 1914). Later, it became an ideology used to describe American society. Once the relationship between the different components of immigration changed, the rediscovery of the "roots" of the various immigrant communities became evident, as always happens in the second and third generation, as portrayed in the *Roots* (Haley 2016). The most recent studies have then adopted the metaphor of the "salad bowl", where the various pieces are mixed while maintaining their mutual identities (Battisti 2006).

9.3 The Institutional Question

Historically, the large European countries were empires, made up of a plurality of small states, each with its own culture (of which the language was an expression), its laws, suited to its particular social and economic structure, and its territorial organization. This is the medieval configuration, in which each state is composed of fiefdoms and inherited lands, linked to each other through a series of oaths (today we would speak of treaties), which at the top referred to the figure of the emperor or king. One example among many is that of pre-revolutionary France, in which the sovereign ruled 2 kingdoms, 1 principality, 14 duchies, 10 counties, and 36 generalities, as well as 9 exclaves. France is an important example to look at, because the Revolution of 1789 will mark the beginning of the Europe-wide transition to modern nation-states. It will be roughly completed in 1918, with the partition of Austria-Hungary after WW1, the last empire with an administrative structure derived from the feudal period (see Chap. 6). In just 20 years after the revolution, the French destroyed the varied fabric of their society, built up over the centuries as the result of the meeting, always different, between the throne, the people, and the altar. In its place emerged a simplified plot, based on a mathematical logic, which deliberately obscured the variety of reality. An intellectual abstraction that anticipates by more than a century the Soviet experiment in post-Tsarist Russia.

With rationality in power, the revolutionary leaders reshaped France as if it were a theatrical play. A single plot replaced the plurality of scenes, the characters of French people rewritten as stereotypical models. The French of the *république* had a single language, a single flag, a single faith—in the goddess Reason—a single center from which came the light that must guide all citizens, and a single uniform: a perfect organization that allowed Napoleon's attempt at world-dominance. Inspired by the successes of Napoleon, other European monarchies will soon start implementing similar reforms to suppress the existing differences within their domains, even though they had declared the restoration of the old system when they met in Vienna in 1815 to rewrite the geopolitical map of Europe after defeating the French emperor. French enlightenment-inspired territorial reforms enabled later political transformations, such as the unification of Italy. Soon after the fall of Napoleon, the Italian kingdom of Piedmont enacted the administrative unification of its historical districts (Piedmont, Savoy, Sardinia) with the recent purchase of the former republic of Genoa. Such territorial reorganization provided the fiscal base for a program of industrialization and rearmament, which allowed it to successfully conquer the rest of the peninsula between 1859 and 1866. Governed by a Piedmontese elite, the newly established Kingdom of Italy was also organized following the highly centralized French model.

Despite these operations, which were carried out in the name of the, at the time novel, ideology of nationalism, five empires survived in Europe until the beginning of the twentieth century: Great Britain, Germany, Austria-Hungary, Tsarist Russia, and Turkey. WW1 spared only Great Britain and Germany. Germany entered the war with soldiers still carrying the insignia of the old feudal kingdoms; after the war, it was transformed into a federation of republics. Great Britain instead to this day maintains some vestiges of the old feudal order: the full name of the country is "United Kingdom of England, Wales, Scotland, and Northern Ireland." Besides these main units, the UK also includes "crown dependencies", i.e. the Channel Islands—each of which constitutes a sovereign state—as well as the City of London, which enjoys a unique regime of autonomy, all examples of the pre-French revolution territorial order. The death of the other three empires led to the birth of a plurality of states in Central and Eastern Europe. The governments of these new states invoked the right to self-determination of the people, but in reality, they repeated within their territories the authoritarian and homogenizing policy of the nation-states, along the lines of France.

Historical events of the 20th and early twenty-first century showed that the assemblage of culturally and economically diverse territories within a single state, once the imperial framework that made up its requirements was missing, created weak geopolitical bodies, undermined by unavoidable internal and external tensions. The most striking demonstration of this came with the separation of the Czech and Slovak republics. This took place in the absence of external pressure, dissolving a Czechoslovakia that seemed to have resolved its ethnic tensions through the expulsion of the German, Hungarian, and Ruthenian minorities in the years following WW2. The split of Czechoslovakia is a clear sign that not only economic differences but also religious ones (Slovakia is 73% Catholic, while Czechia is 35% atheist) have an importance that should not be underestimated (Statistickyurad 2014; International Religious Freedom Report 2015).

The numerous ethnic and religious conflicts occurring within nation-states in the past hundred year show that the 19th-century project of a state for every nation has proved utopian, as it was based on the assumption—wrong—of the existence of a small number of nations rather substantial in terms of population, which could be brought together simply by assembling them in a container geographically adequate. That is, by enclosing them within a border. In the early 1900s, this was still the case for the Germanic world and parts of the Latin world, but not beyond it. Elsewhere in Europe, especially in Central-Eastern Europe, the norm was a dispersion of peoples in small-scale communities, territorially scattered and enclosed in distinct urban and rural spaces; it was a situation not so different from the contemporary Middle East. It was a territorial puzzle that could not survive the ideology of the nation-state without a vast plan of population movement close to ethnic cleansing, as happened—moreover to an insufficient extent—in the 1940s.

Besides creating the preconditions of conflicts, the division of old European empires in a variety of states of very different sizes became an obstacle to the political management of the

European Union, especially after 2004–2007, when the relatively more fragmented Central and Eastern European states joined the Union. To function well, the EU (or any other political entity with the ambition to manage the whole of Europe) should be articulated not so much on the current sovereign states, but on the nations, in their turn promoted to statehood. This would be an effective "Europe of the regions", marked by a leveling down of the size of the geopolitical players. In essence, the construction of a European state should be preceded by the "dismantling" of the nation-states. Such a process would be questionable for countries like Italy that were born and structured following the French Revolutionary model. However, a few other countries could take the road of dissolution. For example, Spain: although it has blocked the independence referendum of one of its larger regions, Catalonia (Dowling 2017), it did not go through the traumatic territorial and social reorganization of France; it still keeps some of the infrastructures of premodern Europe, more akin to an empire than to a homogeneous nation-state. It is a type of organization ill-suited to the invasive interventionism of modern politics, which disregards the country's intimate nature.

There is a real risk of a conflictual secession of Catalonia, which makes it clear what the consequences of a regional restructuring of Europe might be. If the State is, in concrete terms, the container of an economy, its disaggregation inevitably entails the dismantling of the latter, with the very serious loss of fundamental economies of scale. A disaster, unless it is framed within the changes imposed by globalization, whose mechanisms are already redistributing these economies throughout the world.

Considering the uneasy transition from empires to nation-states, and now to an attempt of unification under the EU banner, federal states seem to have an advantage. In particular, Germany continues to enjoy the dimensional advantages of the empire and the flexibility assured to them by a fragmented territorial structure. However, the rigid structure of the territorial models emerging from the French revolution is only one of the problems facing integration. As discussed earlier in this section, the uneven demographic and territorial sizes of European states are also a challenge. German reunification in 1989 aggravated the already existing imbalances between the dimensions of the European partners, creating overnight an agglomeration of forces—demographic as well as economic—capable of taking over the leadership of the continent. Such exceedingly strong Germany increases the risk of conflicts (and exit from the Union) of other states.

9.4 From the European Communities to the Union

As was to be expected, the Europe of the regions has not been implemented at the political level. Throughout the 1980s and 1990s, there were plans to implement stable and devolved subnational territorial units across the EU, which included cross-border regions; however, the project never took off and was instead substituted by temporary regional agglomerations, finalized to short-term EU funding applications (Johnson 2009). The ground on which institutional planning has been developing has been that of the rapid formation of a governance structure that would bypass the problem. This led, in 1992, to the passage from the European Community to the European Union, the latter conceived as a true "superstate", hence an "empire," with characteristics similar to premodern states discussed in the previous section (see also Battisti 2003). The process took place with great speed because it was necessary to respond to the epochal changes that had taken place at a global level: the end of the Cold War, which freed Europe from the Soviet threat and, after almost half a century, promised to remove both the West and the East from the protection of the two superpowers. This was a historic opportunity, a real "window" to be exploited to bring to fruition the unification of the entire continent.

The opportunity of the entry of a large number of new partners, which resulted in the 'fifth round of enlargement' in 2004 and 2007, opened the basic questions of the project: who is part of it, with what rights/duties, with what instruments of government. These fundamental questions were to be answered by the European Constitution, intended as a founding framework to which all future interstate relations would be linked. The first steps of European integration mainly concerned economic cooperation, without prejudice to the administrative framework of the states and their sovereignty. The transition to a political unit required the definition of political relations, i.e. the rules under which the most significant powers pass from the individual states to the new entity. There were two models under consideration, federation or union. The first consisted of a representative body of the individual states, as is the case, for example, in Germany for the Laender. The second, which prevailed, was instead that of the union, with a single-chamber Parliament.

The European Union is, however, far from the centralized model of the unitary nation-states that followed the French revolutionary model. Instead, it has a broad sharing of powers: the EU Council negotiates and adopts laws, together with the European Parliament (but without the obligation to respect its opinion), based on the proposals of the European Commission; it coordinates the policies of EU countries, elaborates foreign and security policy based on the guidelines of the European Council, signs international agreements, approves the annual budget together with the European Parliament. The European Commission, which is the governing body, is elected by the Parliament, whose decisions it implements. It is the only body entitled to propose European laws. Remarkably, the Parliament —the only body with democratic legitimacy—is therefore deprived of the legislative initiative, which is the prerogative of the Commission (the parliamentarians can "ask" the latter to submit proposals). It prepares the budget, which will be approved by the EU Council; however, it does retain some level of financial approval power over the "multiannual financial framework" and on the annual budget.[2] As for the aforementioned EU Council, it does not have a permanent composition but meets in ten different configurations depending on the subject matter. Even from these meager elements, it is clear that the Union is a cumbersome, non-transparent, and essentially disconnected "bandwagon" from the will of the people.

On the level of concrete institutions, the situation is equally confused. Since the draft constitution has not come into force, there is no common European currency. The result is the coexistence of countries that, by adopting the euro, have handed over their monetary sovereignty to the community level and countries that have retained it on their own. Hence a substantial difference in terms of capacity for action. Another difference, based on previous international agreements, can be found (albeit with a different composition) between the "victorious" countries of World War 2 and the "defeated" ones, regardless of their current political and economic situation. This reflects on the entire EU: for example, there is no common armed force; there is a defense policy, which is complementary to, and not an alternative of, NATO (Zamarripa 2019). These are constraints that cannot be eliminated, which place a serious mortgage on any attempt to progress toward an autonomous, democratic, and efficient state organization. In this situation, European activism, in addition to continuing to address the business world, has privileged a legal harmonization on which the Court of Justice and the European Court of Human Rights supervise. These courts, if they open spaces of freedom concerning national jurisdictions, close others, tending to impose on all Europeans an aseptic vision of the Union that allows imposing everywhere the globalizing, disrespectful of local identities and tradition, Enlightenment-driven vision of humanity conveyed by contemporary culture (Battisti 2000).

9.5 The Weight of History

The inability of the political-administrative elites to understand the multifaceted reality of our continent and the concrete limits that stand in the

[2] In a nutshell, the Council approves the budget at the beginning of the year, before it is spent; at the end of the fiscal year the Parliament approves it was spent correctly.

way of a centralizing/homogenizing design that repeats, adapting them to the times, the action of the French and Soviet revolutionaries, is well known. So far their failure has been particularly evident in the unfortunate management of the economy, but the lack of connection between the chambers of power and the feelings of the people is much deeper and lies in the substantial lack of understanding of what Europe is. Our continent is not only composed of a puzzle of nations that can theoretically be assembled at a table, once their individuality has been denied. The various countries that exist today all carry in their DNA the legacy of a series of supranational organizations in which they have participated for a two-thousand-year history. Organizations that in different eras have represented—at least for some—effective alternatives to the design that today would like to be imposed on them, moreover through economic mechanisms that are often inhuman and blackmailing. This is a series of empires that in the course of their history have given from time to time centrality and dignity to a plurality of subjects, which still feel authorized to play a significant role within continental politics.

The homogenizing project of the EU clashes against the long-term legacies of old imperial structures. These legacies are resilient, and the time passed since the dissolution of empires is probably not enough to erase them. Only 100 years have passed since the dissolution of the Germanic, Austrian, and Russian empires; a little more since the end of the Spanish one. Not even 50 years separate us from the extinction of the Dutch, British, French, and Portuguese empires. Further back in time, considering only the modern period, there were many more empires, including the Danish (dissolved after the Napoleonic wars) and the Polish-Ukrainian one (disappeared in '600). Empires are not a simple sum of states, they are complex constructions that are characterized by their civilization that resists homogenization. Last among all, the Soviet empire in Eastern Europe. A group of countries forced to live together but which have thus learned to cooperate in the cultural, economic, and military fields. It is not necessary to emphasize that in many cases we are dealing with realities that stretched mainly outside Europe. This is where the concept of *Magna Europa* (Great Europe) comes into play: such notion acknowledges, similarly to the international literature on postcolonial studies, but in a more balanced way, less focused on blame, that our continent has left durable influences on a large part of the world, albeit in the different variants developed by each colonial power (Cantoni and Pappalardo 2007). This cultural expansion only came to a halt when it encountered an equally ancient, pervasive, and vital civilization, such as the Islamic and Chinese ones. The last imperial expression of Islam, the Ottoman Empire, also dissolved not even a century ago, and arguably nowadays is recovering from a plurality of poles of aggregation. China is arguably alive in a space very close to the old Qing Dynasty and rapidly reacquiring its prominence.

These considerations could sound like a reinterpretation of historical situations very distant from today and therefore appear to be irrelevant to our ability to understand current reality. Such a perspective appears instead necessary when we are faced with a whole series of events that are radically changing the European scenario. *If only because the political constructions of the past represent a sort of archive of memory —similar to the codified positions of chess games —which is recalled every time geopolitical disputes are reopened.* Consider for example the following issues: Iceland's decision to withdraw its candidacy to the EU, Brexit, Poland's political prominence and its opposition, together with the Baltic States, against Russia, and also Russia's intervention in Ukraine. All these events replicate past imperial geopolitical agendas, such as the British focus on the Atlantic, the old link between Poland and Lithuania and its conflict with Muscovy, and Imperial Russia's westward expansion.

Old imperial legacies do not reflect only in member states' foreign policy agendas. There is also a micro-level resistance against Brussels' attempt to homogenize Europe's laws and values. The results are expressed in domestic laws and international agreements. These have a recognizable geopolitical dimension, in the sense that they follow closely old imperial spaces. Take for example the opposition coming from the peoples and Parliaments to the diktats coming from Brussels on birth control, on the definition of the family, on the devaluation of nationality. Hungary has enshrined in its Constitution the Catholic character of the nation, Croatia has defined marriage as between a man and a woman, and Latvia has banned homosexual marriage. Finally, let us recall the unprecedented affinity of views on the "migrant question" that is emerging among the "Visegrad group" (Hungary, Czech Republic, Slovakia, and Poland), to which Austria, Slovenia, and Croatia have been added. The geography of these initiatives replicates closely the space of the old Hapsburg and Polish-Lithuanian empires. Especially the Hapsburg monarchy rested on the twin powers of the Catholic Church (hence the position on church and marriage) and in the close relationship between the monarchy and its peoples, understood as clearly defined groups rooted in place (hence the hostility against migration). In all these events, therefore, it is possible to find a matrix that is linked to the past geopolitical positions of the countries in question.

9.6 Towards a Pluralistic Europe

If we want to move from an abstract vision to the concrete Europe that we see before our eyes today, the scenario that lies ahead requires us to consider a continent divided into a plurality of areas. On the one hand, we have a European Union that aims to absorb all the neighboring states and has recently extended its hands to Ukraine. On the other, we find the Russian Federation, a Eurasian state whose population is predominantly European, as is its culture; committed to building a Eurasian Union, but with a view to a constructive dialogue with the EU. Laying in between these two major poles lays an area of complex interactions. Within the latter, there is a hardcore represented by the "Eurozone", which is dangerously dividing itself between the northern area, based on the German economy, and the Mediterranean front, which is unable to integrate advantageously with the North. The dispute appears to be serious, but unpredictably, the splits are already appearing along an East–West axis. In fact, in the West, we have Great Britain's painful decision to leave the EU, apparently due to the problem of immigration, but actually due to the British desire to maintain control of their currency and not subject their banks to the policies of the Euro bloc. Also not to be underestimated is the desire to distance themselves from the political-economic tensions that are mounting between Germany and the USA. This decision is based on the hope of somehow bringing what remains of the empire back into play, taking advantage of the still strong ties within the Commonwealth. It is also a part of Magna Europa.

In the East between the EU Western European core and Russia, a bloc is emerging under the shrewd direction of the USA, which seamlessly unites all the countries that have left the Soviet empire (plus Austria, Slovenia, and Croatia) under the banner of the "Three Seas Initiative" (no author 2017). Only apparently geared at containing Russian influences, the actual aim is to deprive Germany of the possibility of integrating the whole of Eastern Europe by recreating what was once *Mitteleuropa*—i.e. the alliance between the German and Austrian empires—and of forging further ties with Russia; a constant in German history that Washington and London have viewed with terror for over a century (see Mackinder 1904). Eastern Europe is therefore being militarized once again—under NATO command—with the excuse of a phantom Russian threat, but following a more complex chess game benefiting Atlantic powers at the expense of both Russia and Germany. The recent outbreak of the Russian-Ukrainian war only apparently disproves this evaluation. Instead, the state of weakness and unpreparedness demonstrated by Russian armed forces in the early days of the conflict comes to confirm it, because it obviously conflicts with the interpretations prevailing in the Western world, which do not take into account of Russian security needs. On this point there is however no shortage of critical voices, included that of Henry Kissinger. Such voices counteract the "attitude of dominant and superficial thought… which considers Russia a 'dangerous mix,' inhabited by semi-Europeans who, in one way or another, belong to the world of otherness with respect to Europe" (Cardini, cited in Sanguigliano 2016: xii). On the contrary, today's Russia seems to want to gather from the dust, almost alone, the flag of European civilization (Battisti 2015). G. Sangiuliano sharply underlines: "Compared to the designs of a liquid society, odorless, colorless, homologated to the single thought, Putin is an obstacle because of his traditional, mystic, identity concept of man and peoples" (ibid.).

In this new contrast between East and West, which sees Russia lined up in defense of its national identity—based on Christianity—of moral principles, of the family, of the "natural" birth, it is important to note a significant fact. The national holiday, which was shifted from the anniversary of the Bolshevik revolution to the popular uprising against the Polish invaders that favored the rise of the Romanovs, backdates the founding of the state from 1917 to 1612. In this perspective, Russia remains today the youngest country among the great historical powers of Europe: against England's millennium of existence, France's eight centuries, and Spain's six, with its four hundred years it seems to be able to reasonably hope for a future, once the aftermath of the Bolshevik seventy years has been overcome (Battisti 2009). Only Italy and Germany, in existence for roughly a century and a half, are younger, but their existence covers too short a period to judge whether they can continue to exist in the future. It should also be considered that Moscow, with its 12.4 million inhabitants (18 in the metropolitan area) is today the only European metropolis of world dimensions while retaining the characteristics of the developed world (World population review 2021). Whether we like it or not, we are faced with a plurality of regional groupings, whose strategies fail to converge around a single project. Whatever the new divisions tearing the continent apart, it is the European Union that represents the most relevant structure at the moment and as such, it deserves special attention.

9.7 A Continent in Disarray

Since the peak of optimism following the enlargement of the EU in 2004 and 2007, the EU has been battered by a series of external and internal crises that have led observers to anticipate a collapse of the Union (Kearns 2018). The global financial crisis of 2008 hit Southern European members states especially hard, lasting in some cases up to the middle of the following decade, and exposing institutional weaknesses of the single currency and conflicts between the South and the more affluent North (Hadjimichalis 2011); wars have erupted along its borders, in Ukraine, Syria, and Lybia, causing massive immigration flows that exposed more conflicts between Eastern and Western member states on the management of migration (Cocco 2017); the COVID crisis put the solidarity among member states in serious question (Ferrera et al. 2021), and the relationship with both its strategic partners—the USA and Russia—is increasingly difficult (see the following chapter). Besides these flashpoints that dominated the news in the second decade of the millennium, there are fundamental problems hidden in plain sight, such as the negative demographic trend—the result of low birth rates—which heralds the inevitable disappearance of all European peoples as political and cultural entities, starting with the most fundamental social unit—the biological family, both nuclear and extended, and the principles and shared values associated with it. Wanting to draw mercilessly the sums, we can rightly speak of a continent in disarray.

Faced with this state of affairs, it is natural to wonder what has gone wrong with European integration, given that Europeans cannot put all the blame on external factors, such as the misdeeds of capitalism, the aggressiveness of Islam, or climate change. These are profound existential challenges that exert very strong influences over all forms of political organizations; however, they do not preclude agency and therefore cannot distract Europeans from the necessary task of self-criticism.

9.8 A Gaze in Retrospective

Today we stand 60 years after the start of the unification process, an event that was hailed as heralding an era of peace and prosperity for the entire continent. At its heart was (and still is) the idea that a spatially bounded socioeconomic space with a common identity called 'Europe' already existed, and that it just needed a new management model. According to such a view, the problem of building the European Union is institutional: it concerns the type of regulation that one wants to introduce to achieve the goals one wants to reach. However, this dominant view does not take into consideration the many attempts to "make Europe" following the outline of the plurinational state created by Charlemagne by the likes of Charles V., Napoleon, Hitler, and Stalin. In today's Europeanist and globalist culture, the blame for these failures is placed on a nationalistic approach to politics. There is some truth in this thesis, but it is a tautology: by nature, political power needs a territorial concentration of consensus, nurtured with the management of the economy that supports standards of living. Ultimately, the various projects of a pan-European imperial space did not come to fruition due to the constant opposition of leaders with strong power bases of their own. There is no reason why the current, pluralistic, and democratic EU should not fail for similar reasons.

The dominant, rationalist, and bureaucratic views of the EU tend to forget the religious and ideological background that allowed the early steps toward the establishment of the Union in the 1950s. Our argument here is that faith added an element of cohesion to the initial steps of the EU project that should be acknowledged, and possibly recovered to meet the challenges of the twenty-first century. It is no coincidence that the three "founding fathers" of today's EU, Konrad Adenauer of Germany, Alcide De Gasperi of Italy, and Robert Schuman of France, were Catholics. Schuman and De Gasperi's faith, in particular, was so deep and action-driven that the Catholic Church has started the process to beatify them (Gagliarducci 2021; Mancini 2016).[3] Their faith directly inspired their political action, including the social and economic policies they promoted. They aspired to a unified European polity which, by combining efficiency and freedom, would ensure economic justice in such a way as to produce better living conditions for all and not just for some. They aimed at overcoming the inevitable disparities that emerge between socioprofessional groups and geographically distinct communities. Their Christian-democratic views supported the consolidation across the original member states of mixed market economies (the Rhine, French, and Italian models) profoundly different from the Anglo-Saxon form of governance, based on the pulverization of shareholders, the separation between ownership and management, and the substantial irresponsibility of both toward society. Through their work, the European Economic Community of the 1950s and 60s became a middle way between the 19th-century capitalism of the Anglo-Saxon school, marked by a Darwinist vision of society and the planned economies of the Soviet Union and its allies.

The initial organizational model of the European Communities provided for the gradual transition to shared management of the main issues of civil coexistence. Their approach was federalist, in which power is shared at all levels and national differences are considered an asset to be protected. The main achievement of the early EEC, the one that determined its success, was the European Common Market (ECM). Unsurprisingly, it immediately clashed with the interests of Great Britain, which chose to stay out of it and build its own alternative economic space, the European Free Trade Area, which later

[3] When awarded, beatification will confer the two men a level of holiness second only to saints (Gagliarducci 2021; Mancini 2016).

failed because of the intrinsic weakness caused by, among other issues, its geographical incoherence.[4]

9.9 The Perspective of the Economy

Bringing together the largest, spatially contiguous, economies of Western Europe (France, Germany, Italy, and the Netherlands), the ECM was immediately successful, both in terms of raising the standard of living among the participants and by proving itself more viable than its competitor, the EFTA. However, the ECM did not fully realize its objectives, limiting itself to functioning as a sort of large free trade area. To be truly common, a market must concentrate within itself most of the trade of each partner. As mentioned above, a functional modern state is, in most cases, the geopolitical container of an economy.[5] Achieving such a result was impossible due to the extra-European exposure of the participants in the ECM, a legacy of colonialism, made permanent by the scarcity of cheap raw materials available on the continent. On top of colonial and later postcolonial trade, ECM members were unavoidably linked with the USA, thanks to the latter industrial strength and the benefits of the Marshall Plan.

The heavy dependence of most European economies on extra-ECM trade strongly conditioned further developments. First, it prevented the creation of a core of strategic, self-sustaining activities within the ECM, which would have guaranteed certain freedom of action and independence to the future EU, because an export-dependent economy inevitably configures a political dependence on the outside. Second, it led to a divergence of interests between member states, since the traditional model of international trade requires that manufacturing exports be paid for with imports of raw materials and agricultural products. Since Europe was initially divided between a strongly industrial North and a predominantly agricultural South, their fundamental interests inevitably conflicted, as the North needed low tariffs to export manufacturing products worldwide and the South needed high tariffs to develop its industries and export agricultural products to the North. Moreover, the competition between manufacturing and agriculture crossed the borders between states, introducing internal gradients that are difficult to eliminate, as demonstrated by the case of Italy's *Mezzogiorno*; hence, the birth of what has been acutely defined as "internal colonies" (Hechter 1972).

Unlike farms, manufacturers produce armaments, and the political decision-makers—instigated by corporate leaders—typically protect the interests of manufacturers and financiers over those of farmers, oftentimes manipulating exchange rates to favor trade in manufacturing and services. Military independence was one of the real reasons for import substitution policies, and, after the neoliberal phase of the 1980s–2000, it is one of the reasons for the return of protectionist policies. Unfortunately, in the 1950s and 60s Europe, the size of national markets was insufficient—quantitatively and qualitatively—to guarantee significant corporate profits. By the 1970s import substitution policies were abandoned in favor of export-oriented policies, which later facilitated the acceptance of Anglo-American style neoliberalism. Through this process, by the 1980s Western Europe had become a subcontinent of factories, all export-oriented, exporting manufactured goods which, inevitably, competed with each other.

From the point of view of a common market, Italy should have bought German cars and sold its oranges and oil in Germany. This is what happened during the Second World War, as a result of the economic blockade imposed by the circumstances. Instead, European manufacturers built capacities largely exceeds the absorption of the markets. Hence, already in the 1980s, Europ started an internal policy aimed at spreadi

[4] Members were Austria, Denmark, Norway, Portugal, Sweden, Switzerland, Great Britain, Iceland, Finland, and Liechtenstein, which are not territorially contiguous and whose economies are not necessarily integrated.

[5] Incidentally, this is the reason why the early studies of globalization questioned the viability of nation-states (Strange 1996), and why geographers have been long preoccupied with notions of de-territorialization and re-territorialization (Behr 2007).

consumerism by all means, legal and illegal, which has radically changed the lifestyle of the population, with very heavy repercussions on the spiritual life which, as mentioned earlier, was an unspoken, but important, element of cohesion underpinning the economic model of the CME.

A further economic problem was the production costs of raw materials, which, given the geographical structure (both physical and human) of our continent, tend to be much higher than outside of it. In the post-WW2 period, this meant that the needs of European manufacturers become antagonistic to those of agricultural and mining producers, who needed high tariffs to survive. Hence the real risk of the disappearance of European agriculture, whose production has been gradually undermined by external imports, which are cheaper because of lower costs of labor, energy, and land, and not always of the same quality. If we still have farmers, it is due to the system of strong subsidies guaranteed by the EU agricultural policy; they are not well-tolerated by corporate leaders, who are always reluctant to pay taxes.

Since the beginning of the EU, therefore, there was a patchwork of serious problems, which could not be solved and led to endemic conflicts among countries and economic sectors. The solution would have been a reduction in production structures coordinated among the various countries, which was not viable. A reduction in a homogeneous measure would have meant pushing the weaker economies of Southern Europe to forced de-industrialization and poverty. On the other hand, a more respectful distribution of the needs of the South would have subjected countries such as France and Germany to even more drastic (and therefore politically unacceptable) reductions of their industrial cores.

Europe's problem of industrial overcapacity was the same that the global economy faced as the process of decolonization advanced and some economies of the 'Third World' began developing. The answer to this seemingly dead-end dilemma was, since the late 1980s, the so-called neoliberalization and globalization. Underpinning both phenomena there was a generalized fragmentation of industrial activities, which have been relocated to various parts of the world. The result is that today everyone sells and buys industrial and agricultural/mining products at the same time. The double illusion was that core industrial countries (the USA, Western Europe, and Japan) would have maintained control of the global system through finance. At the same time, industrial delocalization would have allowed suppliers of raw materials to catch up in terms of development. But this was an illusion: the rise of China and the other fast-developing economies, India, Brazil, and Russia, led to the emergence of new centers of power, the global financial crisis of 2007 has exposed the weaknesses of a system based on finance and led directly to the current tensions between the USA, which is attempting to re-industrialize, and China. Moreover, the current COVID crisis and the Ukrainian war are exposing the weaknesses of 'long' transoceanic and transnational supply chains. The ability of the EU to respond to these challenges is yet to be seen.

9.10 Conclusions

From the various considerations that we have summarized here, it seems to be possible to conclude that the unification of European peoples in a single, great homeland, represents a beautiful dream inevitably destined to clash with a harsh reality of internal divisions, external threats, and social and economic disarray. Everything seems to indicate that Europe, originally born in classical Greece, is destined to follow the trajectory of the latter, which historically, once it reached the peak of its development, was unable to achieve either the political cohesion or the economic strength necessary to withstand first the Macedonian power and then the Roman one (Le Lannou 1979).

First, this chapter discussed how the EU is organized following the rationalistic and bureaucratic principles that originated with the Enlightenment and drove the French Revolution, the Napoleonic Wars, and the subsequent emergence of nation-states. These principles include a rationalized form of territoriality that clashes against the legacies of old empires. Papered over in the heyday of post-Communism and EU enlargement, these legacies re-emerged during the various crises since the 2010s—financial, migration, COVID, and the present war against Russia. These flashpoints tested, and are testing, the institutions of the Union and leading to the re-emergence of both interstates geopolitical tensions and anti-EU domestic policies that reproduce closely the spaces of old empires. These tensions spill beyond the EU space, in a complex interplay between the USA, Germany, and Russia, in which the EU is more a pawn than a protagonist.

Second, even though the EU rhetoric is keen to distinguish its peaceful attempt to unify the continent from earlier, war-driven attempts by Charles V, Napoleon, Hitler, and the like, the EU suffers some of the same problems of these early conquerors. Like them, the rationalistic and bureaucratic Union is proving unable to generate enough consensus to avoid the pockets of organized resistance that ultimately failed the earlier attempts to unify Europe. At the bottom line, a vital Union requires a common identity, respectful of people's spaces, traditions, and religious faith, which the cosmopolitan, aseptic, and boundary-less version of citizen promoted by Brussels cannot substitute.

Third, the EU suffers from an economy structurally non-conducive to integration and unity. Simply put, European economies have an excessive manufacturing capacity and too-expensive agricultural and mining sectors and are too exposed to extra-EU trade to fully benefit from integration. Hence, the sharp conflicts between Northern and Southern EU members after the 2008 crisis, and the inability to achieve a common currency that is truly beneficial to EU citizens. This paper in particular argued that globalization from the 1980s to the present time was an attempt to manage industrial overcapacity by keeping rich nations (including Europeans) in control of outsourced manufacturing via finance. The 2008 and especially the COVID crises have exposed the limits of this system in terms of the emergence of new powerful actors and exposed the weaknesses of long supply chains. In so doing, it is bringing to the fore a notion that scholars and policymakers seemingly forgot: that functioning states are in the end containers of the economy and that the EU is functionally unable to contain its members' economies.

Taken together, all these weaknesses seem to indicate that the current European project goes against its historical and religious legacies, its geography, and its established economic structure. However, politics is the art of the possible, which means one must never lead to a passive attitude in the face of difficulties. Recovery of traditional faiths, in particular, could be a powerful engine to achieve that unity in peace—that is, a profound unity of purpose, which makes people willing to make the necessary sacrifices—which the current rationalistic and bureaucratic EU seems unable to achieve. In the end, European peoples have the agency to let Europe survive—or end like ancient Greece did absorbed into a new imperial structure—in this case the American—in which it could not hope to play an important role, but only to be robbed little by little of its accumulated wealth. Indeed, the agency of Europeans must contend not only with the internal contradictions of the EU but also with the latter positioning in 21st-century global geopolitics, which will be the focus of the next chapter.

References

(2017) USA e Germania, duello per l'Europa [USA and Germany: the duel over Europe] special issue in Limes. Rivista Italiana di Geopolitica 5. https://www.limesonline.com/sommari-rivista/usa-germania-duello-per-leuropa. Access 02 Nov 2021

Agnew J (1994) The territorial trap: the geographical assumptions of international relations theory. Rev Int Polit Econ 1(1):53–80

Agnew J (2017) Globalization and sovereignty: beyond the territorial trap. Rowman & Littlefield

Babbitt I (1924) Democracy and leadership. Houghton Mifflin

Battisti G (2000) Il tramonto di una civiltà? [The dusk of a civilization?]. In: Rassegna Tecnica del Friuli-Venezia Giulia 51(4):34–36

Battisti G (2003) Verso una struttura imperiale? Il riaffacciarsi di vecchi concetti nella prospettiva della nuova Europa [Towards an imperial structure? The re-emergence of old concepts in the new Europe]. Bollettino Della Societa' Geografica Italiana 2:355–374

Battisti G (2006) La società multiculturale, ovvero la fine di un illusione [the multicultural society as the end of an illusion]. In Campione G, Farinelli F, Santoro Lezzi C (eds) Scritti per Alberto Di Blasi, Bologna, Pàtron, 173–180

Battisti G (2009) La rinascita della Russia [the rebirth of Russia]. In: Cencini C, Federzoni L, Menegatti B (eds) Una vita per la Geografia Scritti in ricordo di Piero Dagradi, Bologna, Pàtron, pp 315–324

Battisti G (2011) L'Europa come idea e progetto [Europe as idea and project]. Bollettino Dell'osservatorio Internazionale Cardinale Van Thuan Sulla Dottrina Sociale Della Chiesa 7(1):9–15

Battisti G (2015) La geopolitica delle religioni nell'epoca del ritorno del paganesimo [the geopolitics of religions in the age of the return of paganism]. Bollettino dell'Osservatorio Internazionale Cardinale Van Thuan sulla Dottrina Sociale della Chiesa 11(4):145–151

Behr H (2007) Political territoriality and de-territorialization. Area 39(1):112–115

Biggs M (1999) Putting the state on the map: cartography, territory, and European state formation. Comp Stud Soc Hist 41(2):374–405

Burke E (2009) Reflections on the revolution in France, 1790. Oxford University Press, Oxford

Cantoni G, Pappalardo F (eds) (2007) Magna Europa. L'Europa fuori dall'Europa [Magna Europa. Europe outside of Europe]. D'Ettoris, Crotone

Chabod F (1977) Storia dell'idea di Europa [History of the idea of Europe]. Laterza, Roma-Bari

Cocco E (2017) Where is the European frontier? The Balkan migration crisis and its impact on relations between the EU and the Western Balkans. Eur View 16(2):293–302

Dowling A (2017) The rise of Catalan independence: Spain's territorial crisis. Routledge

Ferrera M, Miró J, Ronchi S (2021) Walking the road together? EU polity maintenance during the COVID-19 crisis. West Euro Polit:1–24

Gagliarducci A (2021) Presto un passo avanti verso la santità per uno dei padri fondatori dell'Europa [soon a step towards sanctity for one of Europe's founding fathers] Acistampa. https://www.acistampa.com/story/presto-un-passo-avanti-verso-la-santita-per-uno-dei-padri-fondatori-delleuropa-16938. Access 08 Nov 2021

Hadjimichalis C (2011) Uneven geographical development and socio-spatial justice and solidarity: European regions after the 2009 financial crisis. Euro Urban Regional Stud 18(3):254–274

Haley A (2016) Roots: the saga of an American family. Hachette UK

Hechter M (1972) Internal colonialism: the Celtic fringe in British National Development. Transaction publishers

International religious freedom report (2015) https://2009-2017.state.gov/j/drl/rls/irf/religiousfreedom/index.htm. Access 31 Oct 2021

Johnson CM (2009) Cross-border regions and territorial restructuring in Central Europe: room for more transboundary space. Euro Urban Regional Stud 16 (2):177–191

Kearns I (2018) Collapse: Europe after the European Union. Biteback Publishing

Kirk R (2001) The conservative mind, from Burke to Santayana, 70th edn. H. Regnery Company, Washington, DC

Le Lannou M (1979) Europa, terra promessa [Europe: the promised land]. Minerva Italica, Bergamo

Lewis MW, Wigen K (1997) The myth of continents. University of California Press

Mackinder H (1904) The geographical pivot of history. Geogr J 23(4):421–444

Mancini M (2016) Riparte la causa di beatificazione di Alcide De Gasperi? [Is the procedure for the beatification of Alcide de Gasperi restarting?] Acistampa. https://www.acistampa.com/story/riparte-la-causa-di-beatificazione-di-alcide-de-gasperi-2420. Access 08 Nov 2021

Moisio S (2018) Towards geopolitical analysis of geoeconomic processes. Geopolitics 23(1):22–29

Sanguigliano (2016) Putin. Vita di uno zar [Putin: the life of a czar]. Mondadori, Milano

Statisticky USR (2014) Statistical office of the Slovak republic. https://slovak.statistics.sk/. Access 31 Oct 2021

Strange S (1996) The retreat of the state: the diffusion of power in the world economy. Cambridge University Press

World population review (2021) Moscow population 2021. https://worldpopulationreview.com/world-cities/moscow-population. Access 02 Nov 2021

Zamarripa E (2019) The permanent structured cooperation in the European Union. Its real potential value. In: Ramirez M, Biziewski J (eds) (2019) Security and defence in Europe. Springer, Cham, pp 87–95

Zangwill I (1914) The melting-pot: drama in four acts. The MacMillan Company, New York

10 A Century of Struggles. A Comparison of Multiple Geopolitical Agendas in Europe, the USA, and Beyond

Abstract

Adopting a straightforward realist approach, this chapter discusses the framework governing international relations among powerful states in the century following World War 1. In so doing, it broadens the analysis carried out in the previous chapters from the European to the transatlantic scale. The focus is on the shifts in hegemony among leading powers, comparing the crisis of the UK in the twentieth century with the situation in the early twenty-first century. Conceptually, hegemony is treated in terms of the evolving interplay of cooperation and competition, centering the analysis of contemporary events within the Western world, as driven by the relationship between the European Union and the USA. Continuing his reflections on how geopolitics is ultimately framed by the need to create consensus around economic practices, the chapter emphasizes the geopolitical role of monetary policy. In so doing, it focuses on the function of the European single currency, the Euro, as a barrier against the seigniorage of the US dollar, shedding light on the role of currencies in the competition among states.

Keywords

International relations · Geopolitics · Historical geography · Realism · Long term analysis

10.1 Introduction: The Repetitiveness of Geopolitical Nodes

Beyond the middle of the 2nd decade of the 3rd millennium A.D., the geopolitical situation on a global level presents notable similarities with those that preceded the outbreak of both World War I and II. In both cases, the world was experiencing a phase of depression following a prolonged period of economic boom. On the eve of WW1 in 1914, the world was faced with the crisis of some of the powers that had dominated the international scene during the *belle époquei*—i.e. the period of strong economic growth and rise in global trade at the dawn of the twentieth century. The crisis had hit first and foremost Great Britain, the global superpower of the nineteenth century, whose difficulties were epitomized by the relative decline of British exports in comparison with an increasingly influential German industry. As for France, its difficulties appeared to be no less than Britain. Both the Austro-Hungarian and Turkish Empires were in a state of advanced dissolution, for internal reasons (albeit different from each other). At the same time, three other empires were growing—the German, the Russian, and the Japanese, not to mention the USA—upsetting a already precarious balance of power.

At the eve of WW2 in 1939, the crisis of European colonial powers was linked with

C. Sellar and G. Battisti, *Geopolitical Perspectives from the Italian Border*,
Historical Geography and Geosciences, https://doi.org/10.1007/978-3-031-26044-5_10

Great Depression and economic crisis in the USA, which in 1918 had become the world's largest economy, albeit without the imperial projection of earlier superpowers. At that time, the only power whose economy was recovering was the Soviet Union; Germany and Japan were facing especially difficult situations, which helps to explain their aggressive foreign policy.

At present, we are witnessing the degradation of the dominant power—the USA—and the generalized crisis—demographic, economic, and political—of the European countries and Japan. On the other hand, China, India, and Brazil are strongly emerging, all countries which—not by chance—have a natural demographic balance in surplus.

10.2 The Problem of Primacy

Historically, the loss of economic supremacy is the fundamental cause of the loss of the political one, which always triggers a race for succession to the role of dominant power. In 1914, the declining power was Great Britain, which in the ensuing war burned most of the resources accumulated during the nineteenth century, opening the way to a period of international chaos. Looking at 20th-century history from the viewpoint of competition for global leadership, the two world wars appear as a single conflict that lasted thirty years, in which the outbreak of WW2 in 1939 marked the beginning of US manifested hegemony.

If in 1914–39 the power in crisis was Great Britain, in the first quarter of the twenty-first century it is the turn of the USA. In both world wars, confrontations with powers that had grown in the shadow of the previous dominant precipitated the crisis: Germany in the first instance, Japan and Italy in the second. Now it is the turn of a confrontation between the USA as the declining hegemon, and Europe and the BRICs, especially China, as countries that owe their growth to trade with the USA.

When both Europe and the BRICs are considered as competitors for the position of hegemon, the balance of forces—economically and demographically—appears largely unfavorable to the USA. On its side, however, the USA has an exorbitant military force, able not only to annihilate all opponents through nuclear weapons but also to fight (it is thought successfully) three major conventional wars at the same time.[1] The fact remains that in a horizon of peace, the economic competition leads "naturally" to a downsizing of the States in the face of the intelligent industriousness manifested on other continents. Ergo, the maintenance of the hegemonic role in the world becomes increasingly difficult, if only because the continuous updating of the military instrument involves investments not sustainable by an economic apparatus in strong contraction.

Hence the increasing number of cries of alarm about the decline of America, alarms which, strangely enough, come more from within (Todd 2001; Wallerstein 2003; Chomsky 2011) than from outside, since in other countries there is a tendency to attribute the uncertainties of US foreign policy to the attitude of the president in office (and, secondarily, to Congress), rather than to realistic considerations of costs/opportunities of the country-system. However, there is also a minority of analysts that offer, rather under the radar, a very different interpretation of the events: instead of lamenting decline or blaming the inconsistencies of various administrations, they speak of the birth of the real "American empire.[2]" According to this view, despite the undoubted crisis, the USA has not yet abdicated its global role and therefore it is pursuing its strategy aimed at reversing, if not the dynamics in place, at least the outcome of events. For a

[1] During normal day-to-day operations, the Joint Force will sustainably compete to: deter aggression in three key regions—the Indo-Pacific, Europe, and Middle East. (Mattis 2018, p. 6)

[2] For the short term, however-and by that I mean the next hundred years-I will argue that the power of the United States is so extraordinarily overpowering, and so deeply embedded in economic, technological, and cultural realities, that the country will continue to emerge through the twenty-first century, even though it will be beset by wars and crises. (Friedmann 2010, p. 28)

long time such strategy was less evident than in the past, less rich of blatant statements,[3] but not less pondered. However, in the last few years, the debate concerning the "Great reset" and the clear leadership assumed in the international arena in the development of vaccines during the COVID pandemic and, most recently, in the Ukrainian war seem to corroborate this interpretation.

10.3 Divide and Rule

On closer inspection, in the geopolitical arena (as in any other type of competition), each country competes with all the others. The ability of the leaders is therefore to, first, build a coalition that outclasses all potential adversaries, and, second, to take over such coalition's direction. The main formal coalitions of the current century are shown in Fig. 10.1. As the current hegemon, the USA is formally seen as a friend and partner by at least two of these coalitions (the EU and the Commonwealth), even without formal participation. Coalition leaders must divide opponents as much as possible, both by directly reducing their influence and by taking away their potential allies. To this end, the leader's priority is the identification of the main adversary, against whom to exercise the actions mentioned above. According to the mainstream interpretation of history, the leader was the USA, and the adversary was the Soviet Union, up until the end of the Cold War. The fall of the adversary opened a situation of uncertainty that has lasted a quarter of a century, which today lets us glimpse the possibility that China will become, one day, a candidate for "geopolitical succession".

For a long time, analysts looked at China as a relatively distant threat, not as a reality to be confronted in the immediate future. In the same way, the stiffening of the Obama administration and its successors toward the Russian Federation, which is bringing us back to a "Cold War" climate, is grafted onto a geostrategic situation not even remotely comparable to the challenge brought for almost half a century by the Soviet Union. The Bush-era 'War on Terror' was the response to a trickle of terrorist actions—almost all of them carried out outside the USA—which never put the country's status at risk. Not even Latin America, the traditional arena that for decades has catalyzed US political–military interventions, seems to attract the Pentagon's attention. Instead, the new, unspoken, but real, adversary is the European Union.

Although this seems shocking to all of us, geopolitical realities place the sum of European countries as the new, hidden, main adversary of the USA. These countries have taken the opportunity of the disappearance of the Soviet threat to accelerate their unification process, intending to create a new geopolitical body of global importance. Strong with an enormous human capital (until 2020 the EU could count on more than 500 million inhabitants against the 335 of the USA, all translatable in terms of potential both productive and market), with a large economy, diversified, technically advanced, and fundamentally healthy.[4] Through the process of political unification and the creation of the common currency, "old Europe" was indeed seriously undermining American supremacy.

In particular, the creation of the euro came firstly to protect Europe from the levy of resources (a real "seigniorage")[5] operated on international transactions through the use of the dollar. Second, an internationally recognized and appreciated currency is the basis for the creation of a financial pole alternative to Wall Street. Third—and even more serious—the presence at the world level of an alternative currency to the

[3] See, for example, Bush Jr.'s affirmation, during the final attack on Saddam Hussein's Iraq, that "the standard of living of Americans is not negotiable"; a rather obscure statement, which seemed inappropriate on the occasion and points towards interpretations of the American commitment that are quite different from the traditional ones.

[4] In 2013 the European Union manages to achieve a positive foreign trade balance in both goods and services.

[5] This is the quid pro quo for the maintenance of international order exercised by the US. The concept is well present in the Anglo-Saxon world since the period of British hegemony in the global economy: "Markets require a policeman, and the policeman of free trade was the Royal Navy" (Offer 1989, p. 502).

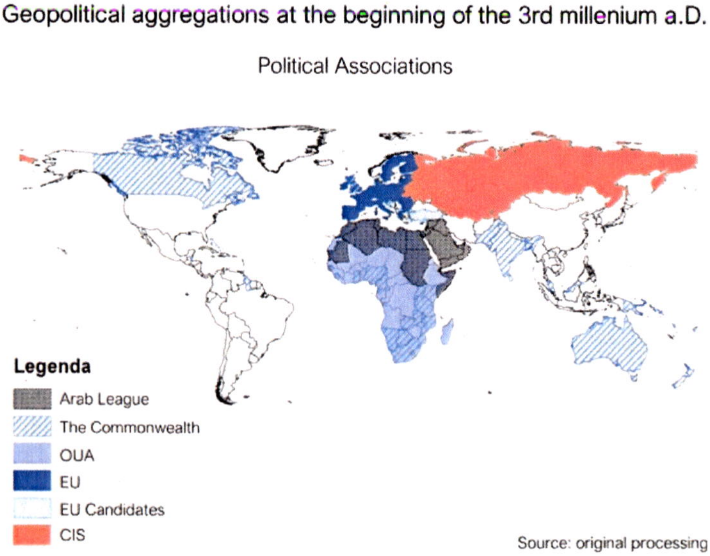

Fig. 10.1 Geopolitical associations. *Source* original elaboration

dollar lays in itself the basis for the collapse of the American economy, which stands only thanks to the advantages deriving from the possession of the universally accepted currency in international trade.[6]

The seigniorage of the US Dollar amounts to a sort of disguised annexation of the other economies to the USA, in the sense that all countries have to commit themselves to support the value of the dollar without having the power to regulate the quantities put on the market. This means that the whole world is indirectly called upon to cover Washington's budget deficits, roughly in proportion to its international trade. This is the situation that has arisen since 1971 when the Nixon administration unilaterally rejected the monetary system established at Bretton Woods in the aftermath of World War II. Against the backdrop of a globalized and financialized economy, this measure can be considered the true birth of the American empire, an empire that is different from all those that have appeared in the past.

[6] Not to mention the advantages enjoyed by certain categories of companies, which are able to secure high profit margins—both export and import—thanks to arbitrage against the national currencies of US trading partners, whose fluctuations can be steered at will by the authorities in Washington.

Given the underlying conflict of interests between the Europeans unifying their economies and the Americans protecting their currency seigniorage, US leaders face the embarrassment to counteract the peaceful development of democratic countries that are not only politically allied but also fundamental trading partners. Not being able to directly exercise the only instrument it has—military force—the USA has opted for a strategy of "peace and security". In practice, it is a policy of encirclement, as we will see in later sections of this paper.

10.4 American Crises

Before discussing the details of the US strategy toward the EU, it is necessary to understand its roots in the systemic crisis of the American economy. The USA entered into a crisis in the first half of the 1960s, following the reconstruction of the European and Japanese economies, which highlighted its competitiveness deficit. An attempt to solve the problem was made through the conflict in Vietnam, which represented a repetition of the Korean experience. On a political level, it was a manifestation of strength, aimed at disciplining both Asian and European allies behind the USA; on an economic level, it was a stimulus policy, aimed at the creation of additional demand in the military-industrial complex. The unsuccessful conclusion of that adventure, however, greatly worsened the crisis, leading to the abandonment of the Bretton Woods financial system, which led to the free float of the US dollar on the international currency market. Since then, there has been a progressive relocation of industries abroad and a concentration of American investments in the financial sector—which lives through uninterrupted expansive and recessive cycles, with consequences that sometimes involve the entire planet.

The financialization of the global economy involved the decoupling of speculative finance from the investment needs of companies, which in turn has made the whole economic system intrinsically fragile and prone to crises. The very nature of contemporary financial crises highlights the relative weakening of US industries. For example, a comparison of the most recent global financial crisis with the Great Depression shows that a crisis of overproduction in American industry caused the financial bubble of 1929, while a crisis of underproduction (caused by overproduction on a global level, which placed the USA in the role of the global consumer) triggered in the 1990s, caused the bubble of 2007–08. Ten years after the beginning of the Great Depression, World War II finally absorbed American productive capacity, allowing the world to overcome the worst economic recession experienced up to that time.

In our time, the downsizing of American industry and the consequent growth of foreign trade deficit indirectly led to the financial bubble of 2007. The bubble spread to the world economy in 2008, first hitting the European and Japanese economies like a sledgehammer and then the former "third world". In retrospect, in 2009 the situation was such that a new general conflict might have broken out 10 years after the bursting of the financial bubble. In a rigidly (and crudely) deterministic vision, the fatal moment should have occurred more or less in 2017, exactly one century after the unlikely success of the Russian revolution changed the reference framework of the world economy for over 70 years. Fortunately, World War 3 did not happen, which invites scholars to investigate the reasons for the escaped danger.

Such investigation should start with the awareness that there is no viable technical solution for economic crises of the magnitude witnessed in 2008 or 1929. More precisely, it is well known to the most experienced journalists, but practically absent from the scientific debate, that the costs connected to an adjustment of the imbalances at the base of such crises are unacceptable to those who control the system. The reason is simple: solving the imbalances would involve a radical transformation of the system, with the consequent redistribution of positions of power. Given the elites' lack of interest in finding solutions that would threaten their positions, one can only wait for the more or less spontaneous solution of the crisis to somehow end up resolving itself more or less spontaneously. The 'spontaneous' solution of 1929 was WW2; no one knows what will come next.

If it is true that the collapse of the stock exchanges has revealed serious distortions in the economy of almost all countries, we must not forget that a conspicuous part of these problems derives from participating in the great casino led by US financial institutions.[7] A speculative orgy whose primary origin lies in the bankrupt state of the American economy. Having collapsed at the end of the 1960s, in the 1990s it was already in conditions of absolute unsustainability, as shown in the budgets of the federal, states, and major civic administrations, all dramatically in red. As if this were not enough, we must add the indebtedness of companies and families to the very heavy public debt, as well as the growing commercial and even financial deficit compared to the rest of the world. After having consumed the capital accumulated in the first half of the century, the USA is now living at the expense of the entire world, substantially exploiting its

[7] It is the "turbo-capitalism" about which Harvey (1981) warned us.

military power.[8] A power that derives from a military-industrial complex that the country is no longer even remotely able to maintain.

Given the US's position of global leadership, restoring the world economy requires restoring the US economy. The second cornerstone of the global system, the Chinese economy, also suffers from major problems, but in this case, the available production potential, as well as the size of the domestic market, are good guarantees that these can be solved. Thus, the USA is the critical element of the system, because solving American problems would entail drastically and permanently reduce defense spending. Per the analysis in the previous paragraph, such a solution is not acceptable by Washington, because it is only thanks to the military that the USA can keep forcing the whole world to feed them. This is more or less what Rome did since the end of the Punic Wars.

The problem is that by now we are reaching the point of no return, as attested by the decision, taken at the end of 2015 by the Federal Reserve, to begin the abandonment of "quantitative easing", the policy of massive monetary expansion pursued since the outbreak of the crisis.[9] However, the Trump administration blocked this decision, continuing to support out-of-control public spending. Despite the "muscular" policy inaugurated in the international arena, foreign accounts have not been restored. At the dawn of the COVID crisis, the situation of the USA has remained unchanged, but the overall picture has worsened both due to the accumulation of deficits and the progress recorded by the rest of the world.

10.5 The Strategic Framework

Against the backdrop of the American crisis, which has spread to Japan and Europe, China, already in 2017 (according to CIA estimates, which correctly rely on purchasing power parity with the US dollar in 2010 as the key indicator),[10] had surpassed the American GDP, with 23.120 billion dollars against 19.519. Even worse, the calculations placed the European Union in second place, with a GDP close to 20.000 billion. Supporting the main argument of this paper—i.e. the EU, and not China, should be regarded as the main competitor of the USA—in the latest update of the 2017 estimates, published in 2019, modified the ranking, placing the EU in first place with 23,339 billion, followed by China (23,210) and the USA (19,490). Considering these data, the Trump administration's obsession with trade balances with the EU and China is more understandable. In reality, as will become clear later, Europe was not able to capitalize on its achievements: estimates for 2019 put China (including Hong Kong) in the lead, with 22,526 billion, followed by the USA with 20,525 and the EU with 21,707. The exit of the UK will cost the latter 14.4% of GDP, which is reduced to 18,587 billion: however, it retains the third position, with a value twice that of India and 4.5 times that of Japan.

Despite its many successes, China's economy still presents significant elements of weakness, due to the strong disparities between the various areas of the country, which compound the dual dependence on foreign markets concerning the import of energy and raw materials on the one hand, and the export of manufactured goods on the other. Theoretically, China has sufficient potential in the domestic market to continue developing in the long term, even in a scenario of reduced trade with Europe and the USA. However, in the short term, China is still vulnerable to trade disruptions: because of the countless linkages between transnational firms,

[8] There are many answers to the question of why the U.S. economy is so strong, but the simplest answer is military power. (Friedman, cit., p. 17)

[9] This term, first used at the beginning of the 2000s in Japan, refers to an inflationary policy pursued by central banks. In contrast to classic "open market operations", liquidity is injected into the system through the purchase of securities issued by private institutions, assuming that these in turn increase the available credit.

[10] China in 2017 stood as the largest economy in the world, surpassing the US in 2014 for the first time in modern history. (www.cia, 2021)

China's first-world industrial apparatus cannot disregard the supply chains that, crossing all the oceans, lead caravans of merchant ships to and from Chinese ports.

The ability to interrupt these caravans is today one of the main missions of the American fleet, which constitutes a permanent threat to the survival of China.[11] There are historical precedents to such use of maritime power, such as the defeats suffered by Germany in the two world wars, largely due to economic strangulation through blockades of its ports by the British fleet. Awareness of vulnerability has pushed the Chinese to take measures that have alarmed the whole world. First of all, they have taken steps to secure their seafront, to protect the regions that have become "the factory of the world". To this end, they are reoccupying many islets in the China Sea, abandoned with the collapse of the Qing Empire, next to which is rising a chain of military bases built on artificial islands. At the same time the Chinese navy, by now the most numerous in the world, has undertaken the construction of a high seas fleet, whose task is not to protect the national coastlines but to control the oceanic routes. From a US standpoint, China's development of a navy with oceanic capabilities constitutes a potential reason for conflict (Laurent 2006, pp. 204–205), as it restricts American ability to intervene in the lives of other countries, which, according to the Wolfowitz doctrine of 1992, named after the then undersecretary of state, must be kept unlimited.[12]

The situation is even more serious if we consider the very difficult situation in which the American economy finds itself (beyond the mythical "recovery" artificially sustained by the emissions without any coverage made by the Federal Reserve). A symptom of this is the new financial bubble that has been inflating on Wall Street (Berg 2015). The impossibility of continuing the policy of deficit spending had already forced the Obama administration to freeze military spending. The freeze was to become a full-fledged budget reduction starting in 2020, just around the time a shift in the dominant position of GDP rankings was expected. It was a difficult choice because China is not the only raising competitor: there is also India, another Asian giant which is talked about as the new *El Dorado*, Brazil, as well as Russia that is finally recovering after the dissolution of the USSR.[13]

While for most countries—especially the former colonies of European powers—commercial relations are necessarily linked to the major sea routes—easily controlled by the US Navy—Russia shares a long land border with the European Union, which is not as easily controllable from the USA. Until 2014, Russia and Europe were increasing trade, due to the natural complementarity of their respective economies. Indirectly, this was part of Europe's gradual reduction in trade with the USA, which does not have much to sell (except high-tech products related to the arms industry and hydrocarbons),[14] nor it can pay for imports, except with increasingly devalued dollars.[15] An economic alliance between Russia and the EU could potentially lead to the birth of a new world superpower, the only one able to keep up with China. A hypothesis that can be read between the lines in the strategies of the EU.[16]

[11] With reserves of 26 million barrels of oil, a domestic production of about 4 million per day and consumption of 14 million, faced with a supply blockage China would go into crisis in less than a week (thewalkingdebt, 2021).

[12] The basis of U.S. foreign policy is a commitment to preventing the rise of powers capable of limiting Washinton's unilateral action. (Roberts 2015)

[13] By this date, Russia is also expected to have modernized nearly three-quarters of its military apparatus (sputniknews.com/military/20151231/1032539—Read 5/1/2016).

[14] In export statistics, the aerospace industry ranks first in 14 states (average value: 19.6%), mining and handling of raw materials in another 14 (average: 32.4); followed by electronics and medical manufacturing in another 6 (21.7% and 11% respectively), motor vehicles and agro-livestock manufacturing in 5 (averages: 19 and 19.2%): source: U.S. Census Bureau data processing, 2019. Overall, traditional manufacturing outperforms advanced sectors.

[15] The share of EU exports to the US has fallen steadily, from 26.4% in 2003 to 16.6% in 2013; 2013 imports touch just 11.6%, compared to 6.9% from Russia (Eurostat data).

[16] To realize this, one need only look at the potential spatial projection of the EU as reflected in the list of

In a new context, such a hypothetical superpower would be an ethically and politically correct reedition of the expansive design of Germany in the two world wars. This is a prospect unacceptable to Washington since it represents a scaled-down version of the heartland concept, which had already raised concerns in the British Empire before 1914.[17] Moreover, as discussed in the earlier sections of this paper, after the 2020s the USA, no longer able to maintain its military superiority, could face the perspective of an economic as well as political collapse. In terms of *realpolitik*, this situation could force the hegemonic power to make drastic decisions, such as, in the extreme case, the start of a conflict to prevent an unfavorable modification of the balance of power.[18] It is no coincidence that Friedman (2010, 121) predicted already in 2009 that the USA will likely face the next global crisis in the late 2020s, in conjunction with the presidential elections 2028 and 2032 (ibid.).

10.6 A Preemptive War

Talking today about the USA starting a preemptive war seems, if not ludicrous, at least 'politically incorrect.' However, judgments of political incorrectness are scientifically meaningless; moreover, most wars in the past have had a preemptive origin. For example, Offer stated in his analysis of the role of food in World War 1: "Great Britain decided several years earlier that it had to intervene in a European war. It was not attacked or invaded. The threat that Britain moved to counter was not immediate. (…) Germany's war of 1914 war [against France and Russia] was described at the time and is often regarded today, as a preemptive war, mounted to prevent a buildup that would culminate in 1916. The British declaration of war [against Germany] in 1914 was a preemptive war of the same type, mounted to prevent a threat that could not materialize in less than two years" (Offer 1989, p. 405).

Similarly, Japanese aggression against China in 1935 was an attempt to prevent the economic recovery of the latter, which, supported by the policy of industrialization initiated by English, American, and French capital, would have destroyed the dream of hegemony on the Asian continent that Japan had pursued for nearly a century. Such tension between China and Japan seems to be recurring: it is surprising that beginning with its economic crisis in the early 1990s, Japan had once again given up its dream of hegemony, this time on a global scale, pursued with farsightedness and determination in the wake of postwar reconstruction. Like in the 1930s, in the past thirty years the industrialization of China has represented the core of Japan's problem. Fortunately, a second, bloody conflict between the two did not happen, and this is undoubtedly a political success. However, at the onset of the 1990s, the continuation of peaceful relations was not an outcome everyone took for granted. At least at the think tank level, the military option was indeed considered (Friedman and Lebhard 1991).

As for World War II, Bandini, in a study entitled *History of a Preemptive War* (Bandini 1971), argued that the guarantee given in haste by France and Great Britain in August 1939 to Poland was aimed at preempting Germany's buildup. Given the material impossibility of bringing any aid to Poland, Bandini argued that the Anglo-French objective was to impose a commercial blockade on Hitler's Third Reich,

Council of Europe member countries (Battisti 2015, pp. 465–466).

[17] On the real influence of the well-known thesis of Mackinder (1904) before 1914, opinions are moreover uncertain (Venier 2004). As for the U.S., this is how G. Friedman expressed himself in a speech to the Council for Foreign Relations: "For the United States, the primordial fear is German capital, German technology, combined with Russian natural resources and Russian labor: it is the only combination that has made the U.S. afraid for centuries" (http://www.rischiocalcolato.it/2015/11/—Read 4/1/2016).

[18] It is worth mentioning that in the matter of nuclear weapons the US military doctrine has included since 1995 the hypothesis of a "first strike" strategy (Chomsky 1999) and that the president is authorized to implement it even without the authorization of Congress. In 2005, the Bush Jr. administration updated the Joint Nuclear Weapons Doctrine by integrating nuclear weapons into conventional combat.

thus repeating the strategy that had defeated the German empire in WW1. Such action was a response to intelligence reporting Italian-German plans concerning the possibility of a war starting in 1942.

Turning to the current situation, it must be noted that today's war is fought through a panoply of instruments much wider than in the past. In particular, as a consequence of the introduction of nuclear weapons, new means of non-conventional conflict have been developed under the umbrella concept of "asymmetrical war," including wider use of espionage, corruption, disinformation, manipulation of elections, political destabilization, sabotage, economic sanctions, targeted assassinations, and many other tools beyond traditional warfare. The seemingly endless war against terrorism, proclaimed by President Bush Jr. in 2001, refers precisely to the large-scale use of such tools.[19]

A careful reading of news of the last decade suggests that the USA has reacted to the multiple strategic challenges outlined above through a well-coordinated series of non-conventional actions beyond the publicized trade war with China, chiefly aimed at destabilizing Europe and preventing stronger ties with Russia; we quote below the most significant ones:

(a) the rupture of mutual economic relations between the EU and Russia by destabilizing Ukraine and pushing the EU to "wrest" the country from Moscow[20];
(b) the encouragement of rifts within the EU by mobilizing in an anti-Russian function the Eastern countries that have recently joined the EU, this attitude is reinforced and at the same time, by assuming the protection of Eastern Europe;
(c) the pursue of a policy of economic strangulation of Russia, through targeted embargoes on exports and imports, aimed at preventing it from regaining economic autonomy;
(d) the attempt at destabilizing Belarus, Russia's last remaining ally on European soil; this is also an attempt to complete the cancelation of Russia's strategic space started with the regime change in Ukraine. From a Russian military point of view, the loss of control over Ukraine and Belarus means that the time available to respond to an attack to the domestic soil has been significantly reduced; from a political point of view, the encouragement given to democratic forces close to home sends a strong positive signal to Russia's opposition[21];
(e) the sanctions and conflicts blocking oil exports of Syria, Iran, Libya, Venezuela force both European countries and Asian allies to redirect their foreign trade toward the USA;
(f) the encouragement, masked as a failed prevention, of conflicts in the Middle East,[22] a traditional trading partner of Europe, is likely to drag the latter into religious conflicts, which have all the characteristics of the clash of civilizations described by Huntington (1996): the massacre of Shiites, Christians, and Yazidis, destruction of historical monuments related to the cultural tradition to which Europe also refers; finally, invasion of refugees and multiplication of terrorist acts attributable to Islamic extremism;
(g) the implementation of a rearmament program, started under the Obama administration, is aimed on one hand at wearing down the vulnerable economies of Russia and Iran

[19] The official name for the *Global War on Terrorism* was *Operation Enduring Freedom*.

[20] In this way they have simultaneously caused Putin's project to build a Eurasian Economic Union, a geopolitical entity capable of dialogue with the EU on equal terms, to fail.

[21] The support for Navalny and the accusations made directly against Putin by President Biden point in this direction.

[22] On the role of the British and American intelligence services (and not only) in what have been improperly baptized "Arab Springs", in the support to Al-Quaeda and IS, as well as in the failure to control Islamic terrorists in Europe, it is difficult to have doubts: see e.g. the hearing of Gen. V. Desportes at the French Senate on 17/12/2014 (http://www.senat.fr/compte-rendu-commission/2—Reading 4/1/2016). Donald Trump himself, then the Republican candidate for the 2016 presidential election, publicly accused, without being denied, the Obama administration of creating ISIS (www.youtube.com/watch?v=OcEMH5Yc—Reading 3/1/2016).

and, on the other hand, at preparing the USA for the decisive confrontation that is looming in the current decade, when Russia will likely have completed the modernization of its military apparatus and China will have achieved the status of great power at the global level.

10.7 The Resurgence of Russia

When all challenges are linked into a coherent strategic framework, the prospect of the USA encircling Russia emerges clearly (Stratfor 2015). Moscow reacted first with a plan for a major restructuring of its military apparatus, the urgency of which had become apparent during the brief conflict with Georgia in 2008. Around that time, Russia began the development of a whole generation of new armaments, which shall significantly reduce the accumulated performance gap with Western countries. Second, on a diplomatic level, Moscow renewed its international activism, bringing it back, if not to the rank of great power, at least to that of a protagonist in global politics. In a nutshell, early 21st-century Russian diplomacy consisted of an extremely opportunistic *realpolitik*, directed toward establishing bilateral agreements with a variety of countries, often led by regimes that are conflicting with each other.

While in Europe, the Russians have maintained control of Belarus by propping up its shaky dictator, in the Middle East they have had to face the Syrian crisis. The potential fall of the Assad regime would have meant the loss of Russia's only military bases abroad, and the construction of a gas pipeline from the Persian Gulf to the Mediterranean Sea. The latter would have meant new competitors against Russian natural gas exporters to the European market. Direct intervention on the ground was therefore the only option to stabilize a strategically crucial front to its advantage. In this endeavor, it allied itself with Iran, while managing to maintain good relations with Israel, which is a nuclear power and an indispensable player in the Middle East. This complex operation appears to mirror the strategy that the USA has successfully pursued for over seventy years, maintaining three different types of alliances with Israel, Saudi Arabia, and Egypt respectively.

The difficult cohabitation with Turkey in the Syrian chessboard was subsequently repeated in Libya, in a triangulation that reinforces the resumption of diplomatic relations with Egypt, which remains a client state of the USA. Moreover, Russia has also begun cautious cooperation with Saudi Arabia, an obligatory interlocutor for a major exporter of hydrocarbons. On the energy policy front, it has also given support to Venezuela, a country under pressure from the USA. And it has also re-established relations with Cuba. In a nutshell, Russia is building alliances wherever it can, creating difficulties to the power system of the USA, which risks being eroded in different areas.

However, Russia's opportunistic alliances are costly in both economic and political terms, and largely of a temporary nature as they are based on a convergence of interests that can vanish at any moment. This is particularly the case of relations with Turkey, but the same can be said about China, a partner that sooner or later Russia will have to reckon with. Trump's attempt to reach a thaw with Moscow was probably based on the awareness of the fragile relations between Moscow and Beijing, displaying a farsighted strategic vision that has been opposed in every way by the "deep state", causing political costs that the former president will still have to bear for a long time. Finally, it should be noted that Putin, notwithstanding the concrete actions described above, has so far maintained a low profile in its public statements, avoiding taking harsh positions toward the USA and the European Union. A cautious position, aimed at safeguarding access to Western markets but also open to possible new scenarios of collaboration because of a probable *redde rationem*—day of reckoning—with China.

10.8 Europe in Disarray

Although it retains a respectable economic potential even after Brexit, the European Union is losing positions compared to the rest of the world. Demographic collapse and consequent aging of the population, downsizing of international trade, and loss of political and military credibility paint the picture of a declining power, a condition that looks even grimmer because the EU has never expressed an autonomous policy. Besides the influences coming from across the Atlantic, the EU's weaknesses are a consequence of the very project devised by the Eurotechnocrats. EU-Europe decided to go beyond the nation-states without creating in itself the "attributes of statehood", generating a "vacuum of authority within and an imbalance of power along its borders" (Kissinger 2015).

The results are there for all to see. After losing control of the Persian Gulf to the USA after the first Gulf War, the EU quickly lost control of the Mediterranean Sea and sees its remaining positions in Africa at risk. From the Baltic to the Mediterranean, in the past decade, all its borders have become insecure. The former satellite countries of the USSR are being transformed into potential battlefields, while the EU is labeling Russia, its main trading partner, as an enemy, albeit it is doing so reluctantly.[23] Moreover, taking into consideration both the difficult relationship with Russia and the explosive situation in the Middle East, especially the worsening of the clash between Saudi Arabia and Iran, we can understand how the EU's traditional sources of oil supply are now strongly at risk.[24]

The only non-conflictual area bordering the EU is the Atlantic Ocean, i.e. the USA, which has recently abandoned the policy followed in recent decades, authorizing again the export of crude oil as well as increasing that of natural gas. The ambition to occupy the role until now held by Russia on the European market appears behind the weak argument of security of supply.

It is easy to foresee that this topic would play a decisive role in the issue of the Transatlantic Trade and Investment Partnership (TTIP), which the Obama administration has tried in vain to establish with European countries.[25] As we know, the terms of the treaty would have entailed the definitive annexation of the latter to the US economy, moreover with the substantial emptying of the EU's productive capacity.[26] Only once the annexation of the entire OECD area has been completed could the USA calmly face the China-Russia axis (possibly extended to India),[27] to prevent the latter from seeking control over the rest of Asia and Africa.[28] Close cooperation on both sides of the Atlantic would compromise these ambitions, perhaps making the use of force unnecessary. Europe's choices could possibly lead to the outbreak of a new world war. In such a context, particular attention should be paid to the "flight" forward of Chancellor Merkel.

[23] The political logic of the EU in dealing with Russia (and China) is clear: "China and Russia are now undermining the international order from within the system by exploiting its benefits while simultaneously undercutting its principles and "rules of the road."" (Mattis 2018, p. 2).

[24] As discussed in an earlier paper, the fundamental element of weakness in European construction lies in energy supply (Battisti 2006).

[25] Alongside the similar *Trans-Pacific Partnership Agreement* (TTP).

[26] The clauses under negotiations were kept secret even from the members of the Parliaments of both the EU and the USA, who would be called upon to approve the text of the treaty without any possibility of modification.

[27] The central challenge to U.S. prosperity and security
> is the reemergence of long-term, strategic competition by what the National Security Strategy classifies as revisionist powers. It is increasingly clear that China and Russia want to shape a world consistent with their authoritarian model—gaining veto authority over other nations' economic, diplomatic, and security decisions. (Mattis 2018, p. 2)

[28] After 5 years of efforts, in October 2015 the US, Japan and Australia had managed to convince 8 other countries bordering the Pacific to sign the TTP, to achieve a free trade area with a clearly anti-Chinese function. Two years later, in a decision that the US will regret, the Trump presidency will withdraw from the initiative. In November 2020, the Regional Comprehensive Economic Partnership (RCEP), the largest free trade agreement ever signed, will be born in its place, with 7 more countries joining the other 11 TTP members. In addition to China and South Korea, the agreement now brings together all 10 ASEAN partners, upsetting the geopolitical balance of the area.

During her long mandate, she tried to safeguard the German economy from the "mortal embrace" with the USA, while at the same time opening up the Chinese market to her companies, which will become the main commercial outlet of the planet.

In this already delicate context, it is important to underline the further, very serious weakening of the EU determined by the unexpected (and problematic) exit of the UK. This epoch-making event has struck at the heart of the European project, which has been deprived in one fell swoop of the largest financial center, the most conspicuous hydrocarbon reserves, a highly effective military apparatus, as well as its real oceanic projection. These are all strategic factors of capital importance both in peace and in war. As if that were not enough, the regained freedom of action of London, introducing a new variable on the geopolitical chessboard, will result in a specular reduction of options available to Brussels.

The divergence of interests between the two sides of the channel is also manifested in the question of relations with Chinadiscussed in the next section. In 2019, total trade amounted to $560 billion (roughly as much as between China and the USA), with a negative balance for the EU of $164 billion (Eurostat data). On this issue, as well as that of relations with Russia, the very survival of the EU is at stake today, as at present the Union represents an essentially economic power. The future will therefore tell whether it can survive in the absence of a unified policy at the international level.

10.9 The Many Problems of an Emerging Superpower

As rapid as the development of the Chinese economy and society has been, such a tumultuous process had to leave large pockets of underdevelopment within the immense country, accentuating preexisting regional disparities. As always happens, the arrival of wealth also triggers the so-called revolution of rising expectations. This has made the social fabric of the regions left behind even more fragile, posing obvious problems for government authorities that are compounded by the sheer size of the country.

China is not a monolithic reality, but rather a set of different worlds, whose coexistence is not always easy. Its most acute problems tend to occur in peripheral areas, settled by culturally and ethnically diverse populations, sometimes similar to those in neighboring countries. The combination of Muslim majority populations with strong cultural connections with Central Asia with the threat coming from radical Islam explains the regime's harsh hand in the province of Xinjiang, which is understandably raising so much criticism in the Western world. Moreover, Xinjiang is not the only human rights-related concern for the leadership in Beijing.

The Chinese leadership's fear of any form of organization that is not rigidly controlled by the single party finds its logic in historical precedents when internal differences triggered bloody and prolonged conflicts, as well as the fear of encirclement and invasion that has characterized the entire twentieth century. Such fear explains the campaigns of 'sinicization' of ethnic minorities, the repulsion of the multiparty system, as in the case of Hong Kong, and the claim to rigidly control all religious groups. Paradoxically, these actions of force are evidence of the fragility of the political system that disturbs the sleep of the rulers, who fear that the great achievements of the last thirty years may be compromised.

Faced with the progressive, probably unavoidable slowdown in the pace of development and the prospect of a progressive closure of its commercial outlets, the Chinese government has launched the Belt and Road Initiative (www.beltroad.initiative), which is China's model of autonomous penetration into global markets. The project has met with a certain degree of success in Asia, Africa, and Russia, but in Europe, which should be its terminal, the reception has generally remained cold.[29] The evolution of the internal Chinese situation—the repression of the Islamic

[29] Greece, improperly devastated by European finance, has opened its ports, some initiative is present in Germany, but the Italian attempt to direct Chinese investments to a strategic port such as Trieste in function of central European markets, was promptly foiled.

population in Xinjiang, the subjugation of all religious communities, and the "normalization" of Hong Kong—seems to pose an insurmountable obstacle to the improvement of EU-China relations. Yet on December 30, 2020, an agreement was reached on the Comprehensive Agreement on Investment (CAI), a bilateral investment **treaty** that replaces 26 preexisting agreements and opens the Chinese market to companies from EU member countries.

The signing took place shortly after the conclusion of another important trade agreement, the Regional Comprehensive Economic Partnership (see footnote 28). The CAI drew strong criticism from outside the EU. Particularly harsh has been the last British governor of Hong Kong, who accuses the EU of pursuing only its mercantile interests, to which it sacrifices human rights and respect for treaties. For Lord Patten, "This is an egregious strategic error at a time when President Biden will seek to bring together a front of liberal democracies to confront the bullying of the Chinese Communists and their assault on international norms. We should not be trying to contain China but to force the Chinese Communist Party [into compliance]. Europe is a joke" (Grotti 2021).[30] For China, instead, it represents an opening of credit capable of thwarting the possibilities of containing its expansion. There would therefore be all the premises for a disarticulation of the "American system", which would open a power vacuum fraught with heavy unknowns.

In practice, the agreement must be approved by the European Parliament, which is by no means a foregone conclusion. After clashing with Russia on the issue of human rights and democracy, it will not be easy for the EU to adopt a different attitude from China, especially after Beijing's countersanctions.[31] Even if it wanted to do so, the clash between Chinese and US diplomacy in the Anchorage meeting (Toosi 2021) represents a precedent destined to strongly influence Brussels' choices.

While CAI was being negotiated, in 2020 the crisis of Hong Kong exploded. Hong Kong's semi-independence relied on its wealth; however, the development of the rest of China has seen its economic weight reduced, so that the city is no longer able to maintain a special status, which has left room for foreign interests in possible conflict with those of Beijing. Therefore, the Chinese government is terminating early the special status agreed with the UK when it returned Hong Kong to China in 1997. The change took place in a traumatic way, with a popular revolt that was repressed with determination and in a professional way, with minimum bloodshed. The conclusion of this chapter of colonial history, however, required a price: the end of the dream of a unified China capable of bringing together several political systems. This was the end of Chinese hopes to reunite Taiwan with the People's Republic by peaceful means. Simply put, Taiwan cannot accept the prospect of being absorbed by a world rigidly controlled by the single party, where even religious freedom is limited.

At present, tensions are rising again around this island-state, whose international space had slowly been reduced to zero. The new Biden administration has maintained an attitude no less firm than that of its predecessor, even though it is aware that Beijing is unwilling to give up a historic objective. The prospect of Taiwan declaring its independence is closer than ever. If it happens, it will be very difficult for Beijing to suffer the affront and renounce to use force to take its province back: an ideal *casus belli* for the powers—in the West and elsewhere—that aim at stopping the rise of China.

[30] This is the point of view of London, which at this juncture is beginning to taste the bitter fruits of Brexit. In fact, it finds itself unable to directly influence the attitude of the EU towards China and at the same time, following the annexation of Hong Kong, it has lost the main card to play with respect to Beijing.

[31] These target 10 MEPs, the Subcommittee on Human Rights, and three other EU bodies, "who seriously damage China's sovereignty and interests and maliciously spread lies and misinformation" (Oliari 2021).

10.10 Escalating the Crisis

Despite American pressure, with the veto of Russia and China and the abstention of European countries, in October 2020 the UN Security Council did not renew sanctions for arms sales to Iran. Washington had to fall back on unilateral measures, against an emerging alliance between China, Russia, and Iran. After the Security Council vote, Iran emphasized with pride how it brought together not just countries but three great millenary civilizations (Gamandiy-Egorov 2020). Iran's stance explains the Eurasian vision that underpins the opposition to American hegemony.

Relations between Russia and the EU have reached an all-time low, as a consequence of an uninterrupted series of pressures (sanctions for Crimea and Donbas, threats to German companies for the Nort Stream 2 gas pipeline, interference in Russian political life through support to the blogger Navalny involving European diplomats). Foreign Minister Lavrov spoke clearly to the representative of European diplomacy Josep Borrell: "We are ready for a complete break with the European Union in case it introduces sanctions that would seriously damage our economy. We do not want to isolate ourselves, but we must be prepared: if you want peace, prepare for war" (ria.ru).

Faced with reactions from Brussels, Lavrov told his Chinese counterpart Wang Yi that Russia suspended relations with the European Union. "There are no relations with the European Union as an organization. The entire infrastructure of these relations has been destroyed by the unilateral decisions of Brussels" (Oliari 2021). In this climate of growing global hostility, the Chinese-American diplomatic summit in Anchorage took place (Toosi 2021).

On that occasion, Chinese diplomats threatened to take "decisive action" against America's interference, which is blamed for a "Cold War" mentality. Given that every escalation of conflict involves a plurality of initiatives, the EU will suffer first in the current situation. Russia has expelled three European diplomats who had participated in the demonstrations in support of Navalny, banned by the Kremlin; in the meanwhile, China's anti-EU measures have been much harder and explicit. Chinese statements in Anchorage have already sent a clear message, addressed to the international establishment that coordinates the "war of sanctions".

The European Union is therefore warned, on both sides, that the hour of irrevocable choices has come: either it remains anchored to the sphere of interests headed by Washington, or it moves in favor of the third-world coalition led by the emerging powers that call for the end of Western hegemony. In either case, it is destined to play the part of the shard among the iron pots in what promises to be an epochal clash.

In the immediate term, the most likely outcome appears to be the end of globalization (Battisti 2019), with the emergence of two geopolitical blocs competing for world hegemony, i.e. for markets and sources of raw materials. Eventually, these blocs will possibly emerge out of the current geopolitical aggregations, shown in Fig. 10.2. A new 'Eastern bloc' is rapidly forming, in reaction to the measures gradually adopted by the West. The alliance with Russia, Turkmenistan, and Iran guarantees China energy autonomy that cannot be threatened by the West. Faced with the blockade of currency transactions implemented against Iran, Syria, and Venezuela, Russia and China have equipped themselves with autonomous electronic payment systems—called MIR and CHIPS, respectively—as an alternative to the western SWIFT and work is underway to unify the two systems. Attempts to prohibit the use of the dollar in international payments are now pushing all threatened countries to seek substitute means of payment. This could favor the Euro, but in the meantime, the yuan, ruble, and Indian rupee are becoming international currencies and gold is experiencing a renaissance, increasingly replacing the dollar in the reserves of central banks. Russia and Iran, oil-exporting countries, are openly advocating a dedollarization of the world, tantamount to a declaration of war against the

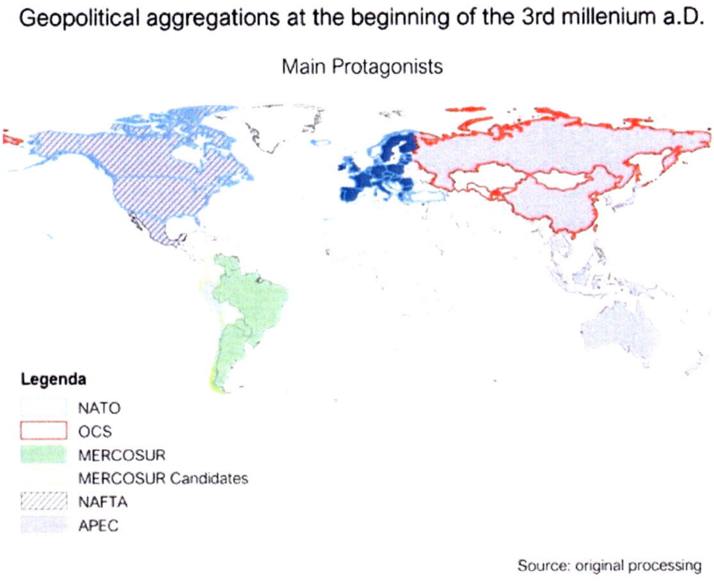

Fig. 10.2 Geopolitical protagonists. *Source* original elaboration

USA.[32] Finally, to protect itself against cyber warfare and component blockade, Russia is replacing all American-made hardware in government agencies with Chinese materials and is preparing to adopt Huawei's 5G system.

10.11 Conclusions

At a time when President Biden is trying to bring American foreign policy back on the tracks of multilateralism, China, Russia, North Korea, and Iran have founded a "Group of Friends in Defense of the Charter of the United Nations", which intends to push for a much more radical change of course.[33] This is not a partisan view; it also finds support in the USA. In his latest essay, Kissinger (2015) no longer talks about democracy but argues that universal values do not reside in a specific form of government to be imposed at all costs. The challenge for America will then lie in overcoming the contradiction between "the celebration of universal principles and the recognition of the reality of the different histories, cultures, and security vision of other regions." The one who had been the architect of the US rapprochement with China saw the situation and wanted to issue a warning. All indications are that we are in the preparatory phase of a global conflict, which for the moment, as denounced by Pope Francis—the only major world leader who has exposed himself on the subject—is being fought "in pieces" (Ansaldo 2014).

Unexpectedly, the beginning of the COVID-19 pandemic in the spring of 2020 changed the strategic picture, leading to an unprecedented "Worldwide lockdown". A sort of "war on the virus" was triggered, which focused the attention of the masses all over the planet. To cope with

[32] From 2015 to 2020, dollar transactions between Russia and China fell from 90 to 46%. In Moscow's foreign exchange reserves, the dollar fell below 50% for the first time.

[33] The other founding countries are Algeria, Angola, Belarus, Bolivia, Cambodia, Cuba, Eritrea, Laos, Nicaragua, Saint Vincent and the Grenadines, Syria and Venezuela, as well as Palestinian representation. A concept note was released stating: "the world is seeing a growing resort to unilateralism, marked by isolationist and arbitrary actions, including the imposition of unilateral coercive measures or the withdrawal from landmark agreements and multilateral institutions, as well as by attempts to undermine critical efforts to tackle common and global challenges" (Nichols 2021).

this emergency many countries adopted a sort of "militarization" of society that paralyzed the normal course of daily life. The response to the virus caused the ruin of most services (from passenger transport to tourism and retail trade), especially in industrialized countries. In a very short time, the world economy has been disrupted: the biomedical industry has become exceptionally important in an instant, while communications and commerce have been massively re-routed on the web. In less than one year, these changes opened up extraordinary opportunities for a radical transformation of the global production structure through the acceleration of innovations in the IT and energy fields.

As we are writing this chapter, these changes are occurring to the detriment of a large number of labor-intensive activities, whose accelerated obsolescence will likely create enormous pockets of unemployment, triggering profound political consequences. As mentioned in the opening of this chapter, the solution to epochal crises involves paying very high prices. If global leaders will succeed in managing the events, technological competition, under the banner of a great world reset (Mueller 2020), could replace the generalized clash that seems to be approaching. There is a real possibility that the global economic situation could offer the conditions for a return to multilateralism (www.ispionline.it). However, such a shift will not make the contrasts disappear.

The even more unexpected entry of Russian Army into Ukraine comes as a final act of the complex itinerary that we have summarized so far. To recognize this conflict as the first step of a long-awaited global confrontation is all too easy. Without pretending to foresee the future, at the present time (June 2022) it seems reasonable to reach some kinds of conclusion. First, the global system has resisted much longer than it did in the Thirties: this gave us five further, precious years of peace. Second, this time has been spent in the development of a global economy restructuring program that goes by the name of "Great Reset". It is still uncertain whether this project, which is dividing peoples and nations, will be implemented. Fourth, there are currently two conflicting narratives concerning the responsibilities for the outbreak of this war, one backing the US-led globalist worldview, the other the security-driven, civilizational defense of Russia. Finally, the determination shown by the European Union in taking up defense of Ukraine, paying the price to subordinate its international trade interests to the political will of Washington, resolves the conflict between the two sides of the Atlantic.

I am not convinced that the EU's painful political decision will be enough to prevent both the spread of the conflict and the emergence of much more dangerous tensions in the future.

References

Ansaldo M (2014) La terza guerra mondiale è già iniziata. La Repubblica 18(8):2014

Bandini F (1971) Tecnica della sconfitta. Longanesi, Milano

Battisti G (2006) „L'Europa che non sarà", Scritti in onore di Roberto Bernardi, a cura di S. Salgaro, Bologna, Pàtron, vol I, pp 439–446

Battisti G (2015) The identity of the European Union as outlined in the constitutional project. Verso unnuovo paradigma geopolitico. Studi in onore di Gianfranco Lizza, a cura di M. Marconi e P. Sellari, Roma, Aracne, pp 457–468

Battisti G (2019) Fine della globalizzazione o tramonto dell'Occidente? Boll Soc Geogr It. Sr XIV, II, 1:129–138

Berg T (2015) Quicksilver markets, OFR Brief Series 15–02, 17 Mar

Chomsky N (1999) The new military humanism: lessons from Kosovo. Pluto Press, London

Chomsky N (2011) American decline. Causes and consequences. Al-akhbar English, 24 Aug

Friedman G (2010) The next 100 years. A forecast for the 21st century. Anchor Books, New York

Friedman G, Lebhard M (1991) The coming war with Japan. Martin's Press, New York, St

Gamandiy-Egorov M (2020) Iran-Cina-Russia: nuova tappa per l'alleanza a tre? https://parstoday.com/it/news/world-i228035-iran_cina_russia_nuova_tappa_per_l'alleanza_a_tre, retrieved 22 Oct 2020

Grotti L (2021) Mentre la Cina reprime Hong Kong violando un trattato internazionale, l'Ue trova un accordo commerciale con Pechino! Tempi. https://www.tempi.it/laccordo-commerciale-tra-ue-e-cina-e-un-errore-strategico-madornale/, 12 Jan 2021

Harvey G (1981) The condition of postmodernity: an enquiry into the origins of cultural change. Wiley-Blackwell, Hoboken, N.J.

Huntington SP (1996) The clash of civilizations and the remaking of world order. Simon & Schuster, New York

https://www.beltroad-initiative.com/belt-and-road/, retrieved 15 Mar 2021

https://www.cia.gov/the-world-factbook/field/real-gdp-purchasing-power-parity/country-comparison/, edition 11 Mar 2021

https://www.ispionline.it/it/pubblicazione/global-watch-coronavirus-speciale-geoeconomia-n51-29978/, 9 Apr 2021

https://ria.ru/20210212/lavrov-1597157215.html, 12 Feb 2021

https://thewalkingdebt.org/2021/03/15/la-questione-energetica-cinese-la-grande-sete-di-petrolio/, 15 Mar 2021

Kissinger H (2015) World order: reflections on the characters of nations and the course of history. Penguin Books, London

Laurent L (2006) La verità nascosta sul petrolio. Un'inchiesta esplosiva sul "sangue del mondo'. Nuovi mondi Media, Bologna

Mackinder H (1904) The geographical pivot of history. Geogr J 23(1904):421–437

Mattis J (2018) Summary of the 2018 national defense strategy of the United States of America. Sharpening the American Military's Competitive Edge. https://dod.defense.gov/Portals/1/Documents/pubs/2018-National-Defense-Strategy-Summary.pdf

Mueller AP (2020) From lockdowns to "the great reset", Mises Wire. https://mises.org/wire/lockdowns-great-reset, 8 Jan 2020

Nichols M (2021) China, Iran, North Korea seek support at U.N. to push back against unilateral force, sanctions. https://www.reuters.com/article/china-usa-un-int/china-iran-north-korea-seek-support-at-u-n-to-push-back-against-unilateral-force-sanctions-idUSKBN2B336H, 12 Mar, retrieved 18/3/2021

Offer A (1989) The first world war: an agrarian interpretation. Clarendon Press, Oxford

Oliari E (2021) L'UE sanziona tutti. Lavrov, non abbiamo più rapporti. https://www.notiziegeopolitiche.net/lue-sanziona-tutti-e-lavrov-non-abbiamo-piu-rapporti/, retrieved 23 Mar 2021

Roberts PC (2015) War is on the horizon: is it too late to stop it?

Stratfor (2015) The west hems in Russia Little by Little. AS the U.S. and NATO presence in Eastern Europe grows, Turkey and Turkmenistan could assist in the attempt to encircle Russia. https://worldview.stratfor.com/article/west-hems-russia-little-little, retrieved 30 Mar 2021

Todd E (2001) After the empire: the breakdown of the American order. Columbia University Press. New York

Toosi N (2021) China and U.S. open Alaska meeting with undiplomatic war of words. https://www.politico.com/news/2021/03/18/china-us-alaska-meeting-undiplomatic-477118, 18 Mar 2021

Venier P (2004) The geographical pivot of history and the early 20th century geopolitical culture. Geograph J 170:330–336

Wallerstein I (2003) The decline of American power. The U.S. in a chaotic world. The New Press, New York and London

11 Cycles of Geopolitical (Dis)order, as Determined by Interactions Between Spatial Systems. A Theoretical Model of the Systemic Drivers of Geopolitics

Abstract

This conclusive chapter ties together the earlier reflections on the interactions between historical geographies, economics-framed geopolitics, and globalization to analyze the systemic nature of interstate conflicts leading to globalization since the early modern period. It starts with a nod to realism by treating legally recognized territorial states as units in a global system, and by defining geopolitics as the spatial relations among such units. However, the chapter also takes stock of a variety of underlying systems (cultural, economic, and even institutional) that tend to flow across states' borders. States are the most important element in this 'system of systems,' because the political sphere maintains the consensus necessary for the other systems to function. They do so by managing changes in the other systems while maintaining social peace; therefore, political crises at all scales—from micro-level unrests to widespread war—are indicators of systemic transformations that states are not able to manage. Such an approach enables interpreting the long-term development of globalization in terms of equilibrium, a concept borrowed from economics. In Battisti's view, spatial expansions and contractions of multiple systems have shaped globalization. Relative equilibrium among systems characterizes periods of peace in specific areas, while disequilibrium is the root of wars and conflicts. The efforts to achieve equilibrium in the face of constant change spearheaded the evolution of the political system into modern territorial states and beyond.

Keywords

Globalization · Systemic analysis · Interstate conflicts · Early modern period · Long term analysis · Equilibrium · Geopolitics

11.1 Introduction: A Systemic Approach to the Geopolitics of Conflicts

This final chapter fully develops the idea that, if geography should be considered an organic and unitary discipline across its subfield, then discussions of geopolitics should be keen to include the geographies of a multiplicity of human experiences. Specifically, this chapter looks at interstate conflicts and struggles as the result of a constantly evolving balance between different systems emerging from people's various social activities across space. In so doing, it defines as geopolitical order the geographical expression of spatial relations within a territorial system composed of internationally recognized state units. As such, it is subject to continuous dynamics, which propagate between the different thematic areas—cultural, economic, and institutional—

C. Sellar and G. Battisti, *Geopolitical Perspectives from the Italian Border*,
Historical Geography and Geosciences, https://doi.org/10.1007/978-3-031-26044-5_11

exceeding the spaces delineated by individual borders. Of the countless relationships that can be traced, the most significant appears to be those concerning the most important sectoral subsystems (political, economic, and territorial). A further clarification must be made: with the modern era, economies-states that were closed to the outside world gradually disappeared. Thus, the road to the modern world from the European explorations of the sixteenth century to the end of colonialism after WW2 can be described as the progressive establishment of the global geopolitical order, marked by the development of cartography as mean to acquire knowledge supporting political decision-making.

Within each state, there are continuous transformations in economic and social structures that follow a variety of trajectories—both in terms of growth, but also decline and crisis. For the most part, social scientists analyze these changes separately—for example, the trajectory of a regional economy is not part of analyses of ecosystem changes. However, this chapter argues that the main role of the state system is the management of changes in all social structures distributed across each state territory. Beyond a critical threshold, which it is often impossible to determine in advance, the structure of the state system is unable to incorporate all the changes smoothly and experiences a crisis. This chapter also argues that the state system is not just one among many social and economic systems composing human societies on Earth. Instead, it has the role of a keystone, because it oversees the functioning of all other systems. When the rate of changes among the various systems is manageable, the state system is in equilibrium: the relations among governments are stable, and the alliances and international institutions operate without major conflicts. Instead, major interstate disputes or conflicts and challenges to the legitimacy of international institutions and alliances are indicators of profound changes across systems.

The keystone position of the state system is both a strength and a weakness since it forces states to adapt—by modifying both their structure and their functions—to the changes registered by the other subsystems, challenging both domestic policies and the role of territorial subdivisions within states, and the relations among states, regardless of the presence of a common border. Most commonly, crises within states arise from transformations in two or more interconnected social systems. For example, in the past thirty years several European states experienced more or less serious threats of secession: in the 1990s a political movement argued for the secession of the Po Valley from Italy, in 2017 Catalonia held an independence referendum to leave Spain, Northern Ireland experienced a civil war concerning its union with England, Czechoslovakia peacefully split itself in Czechia and Slovakia, and Yugoslavia broke up in a bloody civil war. All these conflicts have had in common a geographically localized economic dispute, which overlapped with cultural differences rooted over time in the same areas. The respective states were threatened when opinion leaders in both the cultural and economic spheres joined forces to elaborate a common political agenda in conflict with the one developed by the central government.

In turn, crises of the magnitude of a secessionist movement rarely remained entirely contained within a single state and may cause potential threats to international relations. International disputes of various intensities, up to military conflicts may arise when the transformations triggered by the domestic struggles lead to ruptures of geopolitical balances, which are manifested through the abrupt change in interstate relations. From economic/cultural exchange we pass to economic/military confrontation, which is generally, but not always, of short duration.

The dynamism of the various systems leads to shifts between stable geopolitical arrangements, in which the various subsystems are either in equilibrium or in slow, manageable transformation, and unstable ones in which relations are rapidly transformed toward a substantial modification of the equilibrium in the state system. People's instinctive aversion to violence and the habit to focus attention on domestic events lead to the perception of wars as exceptional events, which occur from time to time. Some argued, incorrectly, that the frequency of these events has declined throughout history (Pinker 2012). However, a gaze beyond borders, especially beyond the borders of the relatively stable Western democracies shows that this is, unfortunately, an illusion. On a large scale (especially on a world scale), it must be acknowledged that conflicts of various intensities, up to actual wars follow one another here and there practically without interruption, to the point that the periods in which the whole Earth is war-free are negligible.

11.2 The Role of Conflicts

Wars are continuous because they arise from specific disputes, which are hardly ever settled on a single occasion. Hence, the emergence of cycles of wars, which can be concluded only under two conditions: (a) the decisive prevailing of one of the contenders, (b) the intervention of an external power interested that forces the contenders to suspend at least temporarily their conflict. If we look at the spatial scope considered, wars can be roughly divided into territorial, global, and 'glocal' wars. The first and the last category include the overwhelming majority of wars. By territorial we mean wars that are localized to a single area; by glocal, a conflict that usually starts locally ends up affecting a wider area and involves some global power. Lastly, the notion of global is not exclusive to the planetary scale, as it has become the practice in the common language since the 1990s. In this essay, 'global' means something that involves the whole of a given spatial system, regardless it covers the whole earth or not. Such definition is close to Wallerstein's original formulation of World System Theory, in which he analyzed the emergence of a capitalist system initially confined to Europe and then expanding to the rest of the world (Wallerstein 2015).

In this chapter's definition, the notion of global wars is similar to what historians have called 'systemic wars', which involve the entirety of international relations considered (Midlarsky 1990). These are usually the central events in a cycle of wars that bring together all the actors involved and reshape the international system in its entirety. They redefine (or reaffirm) the center of the system and the hierarchical-functional position of all actors. This chapter defines those as 'global' because they reshape the entirety of the world of the actors involved, as it was the case, for example, of the Peloponnesian War between Sparta and Athens, the Punic Wars between Rome and Carthage, and the Hundred Years War between England and France. Wars that are not systemic (or global) are local, such as, for example, the conflict between Georgia and Russia in 2008, which was in practice contained between the two participants and did not have apparent systemic changes. However, especially long-lasting local wars may attract new actors assuming a "glocal" character, giving the conflict broader implications. In some cases, such events may become the trigger for a global conflagration, such as, for example, the confrontation between the Austrian Empire and Serbia in 1914.

Conceptually, the analysis of conflicts depends on the spatial framework, which provides the overall view that the observer chooses to privilege. Analytically, the geographical analysis of wars and conflicts is part of regional geography, in its systemic meaning (Vallega 1995). Such analyses are necessarily difficult and incomplete since the various systems are embedded in each other in a dual structure: vertical—i.e. the hierarchical relations between the state system and the other subsystems—and horizontal—i.e. on the spatial representation of systems on a map, each subsystem overlapping and sharing fuzzy boundaries. The boundaries are fluid in both vertical and horizontal relations: this is because the different systems cross into each other in an almost unresolvable jumble. It is for this reason that the search for the 'right' boundary' has never led to satisfactory results. The lack of clear boundaries between systems makes it difficult to monitor changes within individual systems, all the more so since the metrics to be considered are both spatial and temporal.

Wars and crises more in general simplify the analysis of the relationships among the various systems because they signal systemic ruptures that might otherwise be overlooked. They are privileged occasions in which to become aware of the real extension of the systems and subsystems at stake. However, the most significant ruptures involve more than individual crises: the most significant changes in the relations among systems involve clashes prolonged over time, in which a plurality of actors constantly enter and exit a state of conflict until the entire global system has reached a new balance.

Treating conflicts as flashpoints of more complex and prolonged systemic realignments leads to the question of periodization, a problem that confronts historians with questions similar to those that trouble geographers concerning regionalization. In practice, it is possible to identify connections between wars, but the effort of identifying precise temporal boundaries between cycles of wars is rarely possible to accomplish. The most evident example is that of the two world wars. The conflict between the countries involved in these tragedies originated well before 1914 and in most cases has not died out to this day. The current differences within the European Union, of which Brexit is just the latest and most glaring example, are ample proof of the persistence of these geopolitical tensions.

Most conflicts lend themselves to multiple interpretations—all equally true and worthy of scientific consideration—as they depend on the perspective from which they are viewed. This perspective can only be correctly based on the specific spatial reference system. Geographically it is a problem of scale, a scale that is measured, however, not in physical terms (which concern cartographers and transport technicians) but in terms of systems. For example, the internal conflict that has persisted in Libya since the ousting of the dictatorship in 2011 has been interpreted very differently within Libya, in Italy, France, Egypt, Israel, Saudi Arabia, the USA, and Russia. The conflicting views of the Libyan crisis show the dangers of focusing on a single systemic (and therefore spatial) level, even the global one (the reference system for the hegemonic power at the time) because this can lead to catastrophe.

Two examples are sufficient to clarify the point. In the fifteenth century, Charles the Bold viewed the Swiss Confederation as a secondary actor in the grand geopolitical game among European powers. His view was consistent with the idea, dominant at the time, that politics was centered on large dynastic kingdoms. Instead, his lack of consideration for a small republican politi destroyed the project to create a new political power in the Burgundian territories finally costing him his life. Had he won, European history may have taken a very different direction, to include a new imperial power in Burgundy, today's border region between France and Germany. In the twentieth century, the USA made a similar mistake by considering Vietnam as a local piece of the global system. However, the local population viewed it as the heart of a regional system encompassing Southeast Asia. Unlike Burgundy, the consequences for the USA were not lethal; however, the defeat in a local war with systemic consequences across Southeast Asia delegitimized the USA, bled it economically, and damaged its political credibility for more than two decades, until the fall of the USSR in 1991 left it as the undisputed superpower.

1991 marked the moment when the concept of symmetrical and asymmetrical wars became a focus of scholarly and policy attention. When passing from a bipolar balance to the hegemony of a single superpower, wars can only be unbalanced in favor of one side. The intervention of the dominant superpower, however, is hardly politically acceptable; in fact, the USA has carried out its recent wars under cover, either in the guise of an international alliance (see Gulf War), peacekeeping missions in Iraq and Afghanistan, or various types of offensives—guerrilla (as in Syria), psychological warfare, cyber warfare, and economic warfare—whose objective is always the destabilization of the adversary.

11.3 Globalization in Waves

In an earlier article, Battisti defined globalization as a process that tends toward the spatial increase of geopolitical systems (Battisti 2011). It is an oriented and direct process, in that there is always the main actor who outlines the strategy, gives the directives, and ultimately bears the brunt of it (even if seeking every opportunity to offload the costs, at least in part, on allies and adversaries). In the colonial expansion of European countries, the list of protagonists is very long, but in every historical period, it is possible to find a hegemonic power that has achieved the most significant share of the advancement. Portugal began, followed by Spain, then the Netherlands, and later France joined the field, followed shortly by Great Britain. Wallerstein focuses his attention on the two Northern European powers, whose economic development takes on the characteristics of modern capitalism, which he defines as capitalism tout-court (Wallerstein 2015).

One of the fundamental ruptures in history is the revolt of the seven protestant provinces of Flanders, which will succeed in asserting their independence from the Hapsburgs after 80 years of wars, thus founding the modern state of the Netherlands. According to Wallerstein (1978), the birth of the Netherlands marked the beginning of the "world-system", an original interpretation of globalization. The model we have proposed in our turn does not recognize such a recent genesis; in fact, it examines, rather than the birth of capitalism, the major transformations of the economy: the agricultural revolution, the industrial revolution (which requires ever-larger markets), and the tertiary or postindustrial revolution, as we like to call it (Battisti 2012).

If we conceive of globalization as a repetitive process, composed of phases different from each other while moving within the broader unification of the planet that comes from the relentless development of the media, a new vision of geopolitical dialectics emerges. In an earlier publication, Battisti distinguished three major phases of globalization (Battisti 2017), which can in turn split into a series of narrower spatiotemporal processes, corresponding to the individual phases of expansion of trade/cultural exchange. We can then speak of "waves of globalization" that move like the waves of the sea, in an alternation of ebbs and flows; which are marked by the enlargements of the area where trade is liberalized and the subsequent autarchic closures of country-systems and their coalitions. The overall structure of the current wave, characterized by the global interconnection of regionalized economies along searoutes, is portrayed in Fig. 11.1.

The ebbs and flows of globalization are shaped by the simultaneous development of production and communication technologies; the first leads to the spatial widening of economic

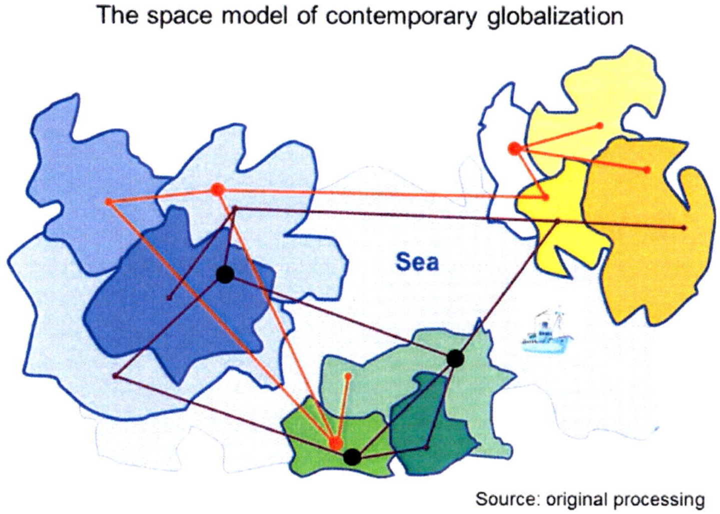

Fig. 11.1 The space model of contemporary globalization. *Source* original elaboration

relationships, while the second enables it. The need for production to find sales in larger areas alternately pushes in two directions. In the beginning—taking advantage of a stable geopolitical order—international agreements were made aimed at the liberalization of trade, as occurred in Europe from the 1860s to the end of WW1. The tumultuous growth of production that followed led to overproduction, which in turn led to widespread protectionist policies. In the timeframe considered, this policy was inaugurated in 1879 by Bismark, during the worldwide "great depression" of 1873–1895, characterized, as today, by strong deflation. The search for new resources and new outlets led point to imperialist policies, which reduced the spaces economically accessible to all. As a result, under the illusion of escaping national protectionism (Aparicio Cabrera 2013), an expanded protectionist system was built, which reduced the freedom of trade. In practice, the expansion of colonial empires in the nineteenth century was an attempt to carve commercial outlets to solve the productive overcapacity of each European power. In so doing, the process of globalization shifted from being unitary and multipolar to being heavily regionalized, consisting of trade occurring for the most part between European metropoles and the colonies, with each colonial empire economically separated by high tariffs and customs duties. In essence, the late nineteenth century was characterized by agglomerations in which each industrial power attempted to build an enlarged economic space in the search for a new balance between production and consumption. Such expansion only delayed the inevitable clash between the major industrial powers, which will eventually lead to WW1 (Mulligan 2017).

The economic consequences of WW1 lingered for a long time in Europe, leading to widespread political instability. On the other side of the Atlantic, the USA experienced an economic boom throughout the 1920s, which came to an abrupt halt in 1930, with the beginning of the Great Depression. The Depression will spread to the whole world because of renewed protectionist policies that each major power implemented in an attempt to protect its domestic economy: these included the system of imperial preferences within the British Empire, the Italian autarchy, the iron clearing policy in Nazi Germany, and many others. Including the subsequent WW2, the stagnation of globalization lasted a total of fifteen years, up to 1945.

Once WW2 ended, the process of globalization resumed with a new spatial dynamic. In the 19th and early twentiethcenturies, each industrial power had attempted to build exclusive economic spaces within its colonial empire; after 1945, the beginning of decolonization led once more toward an organization of the world economy less bounded by borders. However, the beginning of the Cold War led to a split of the fledgling global economic system in two. In a nutshell, from 1945 to 1990 the world experienced two different economic systems, as a consequence of the USSR opposing the creation of a unified geopolitical area on a global scale. It did so by refusing to join the United Nations Relief and Rehabilitation Administration (UNRRA) and the US-led Marshall Plan, aimed at the reconstruction of the countries ravaged by the war. Instead, the USSR led the reconstruction of the Communist countries according to a planned economic model that substantially prolonged in peacetime the characteristics of a war economy. Geographically, the Communist world included a large portion of Europe and Asia: an area rich in natural resources, but suffering from the serious destruction of war. However, the Soviet bloc suffered from the imposition of an illiberal regime and the autarchic closure toward the rest of the world. Moreover, competition with the much wealthier USA led to costly military expenditures: by the 1980s, the USSR spent around 15% of its annual GDP for the military, roughly double the USA (Holtzman 1982). Taken together, illiberalism, scarcity of trade with the rest of the world, and excessive military expenditures explain why it only took four decades since WW2 for the Soviet bloc to collapse and for the world to be reunified under a single economic direction. Its fall was unavoidable, despite some moments in which it seemed that communism was destined to spread everywhere, prevailing not only materially but also intellectually.

11.4 Toward the End of a Wave

The fall of the communist system brought down the geopolitical duopoly, opening the door to the virtual unification of the world under the banner of the capitalist economy. The new phase in the global political economy that started in 1989 was marked by the widespread use of the term globalization, which was used to endorse the mistaken impression that the integration of the world economy is a phenomenon that started with the postmodern era (Fumian 2003). Although not new per se, this most recent wave of globalization brought about a radical reorganization of international relations. In particular, it has led to an international redistribution of labor which, on the one hand, has destroyed a large part of the productive capacity of the former communist countries (think of East Germany, perhaps the most efficient economy in the East, or the downsizing of Soviet military industry) and, on the other hand, has led to the rapid industrialization of China.

Looking at the events from the old bipolar perspective, there has been a sort of exchange of roles between the two largest countries in the former communist world, with China overtaking Russia. From the new, unipolar perspective but what matters most is the scale of China's economic development, able to conjugate the old Communist ideology and Leninist political infrastructure with a leading role in international trade. The growth of China challenges the US hegemony not only on the economic and geopolitical ground but also ideologically, showing that successful capitalist economies and democratic regimes do not necessarily coincide. Moreover, the increasing geoeconomic and geopolitical centrality of China and its East Asian neighbors, together with the weakness of Russia and the relative decline of the West is leading a global system closer to the one that existed until the seventeenth and eighteenth centuries, when China and India were by far the wealthiest countries in the world. Such change is questioning another cornerstone of Western modernist thought—the notion of unlimited linear progress—in favor of older, cyclical views of development (Lizza 2009, 21). The rise of Asia has also had systemic economic consequences since the massive delocalization of productive activities in the countries of the former "third world" has rapidly emptied the old "industrialized world" of productive activities. The transformation has radically affected the heart of the system, namely the UK and the USA, which have gradually abandoned the less profitable and less competitive activities to the satellite countries to focus only on managerial activities: administrative, financial, scientific, communication, military.

Several empires in the distant past had experienced economic transformations similar to those occurring today in the USA and UK. Ancient Rome did it in the aftermath of the Punic Wars; Spain repeated it at a much quicker pace than Rome at the beginning of the modern age. In different historical periods both Rome and Spain, thanks to military force, had built their own "global system" which ended up delegating the production of wealth to treacherous clients and to the peoples they defeated. Today's globalization follows a similar logic, which requires the economic subjugation of the defeated countries through a "soft" formula, which avoids as far as possible the waste of resources resulting from the use of force. In past and present globalizations, the subject countries must be involved in a 'great market' represented by the consumption of the dominant power on its territory and on that of the satellite states (Pollio Salimbeni 1999). In contemporary globalization, the 'great market' is represented by the USA, which has the role of global consumer in a world market it organizes through bilateral and multilateral organizations and agreements. At the strategic level, the consumer role of the USA has also the advantage of directing the subject countries toward non-military productions, which guarantees the control of the most strategically important armaments. The similarities with ancient Rome are striking (Luttwak 1981). Back then, it was an obligatory path for the top of the system, but in the long run, it produced the same deleterious effects that have been recorded many times throughout history (Todd 2003).

Today we have reached the point of no return: the USA must urgently reduce the gigantic deficit in the current account (Battisti 2012), Great Britain must escape the grip of a European Union that would inevitably end up regulating the economy and decide its future. Brexit on the one hand and the wavering between multilateralism and bilateralism in the Obama, Trump, and Biden administrations in the USA seem to indicate an underlying awareness of the vulnerability inherent in the role of the global consumer. Beyond rhetoric, there is a surprising level of continuity between Trump and Biden's administration in certain policies, such as the withdrawal from Afghanistan, the management of immigration, and the stance of China. Amidst the supply chain problems during the COVID crisis, elites in both Washington and London seem to move in the direction of re-industrializing their respective countries. Such policies must necessarily entail an increase in exports and a drastic reduction of imports. A huge task, a commitment whose duration cannot be less than the time it took to empty the two countries of their manufacturing activities—i.e. the 40 years since Reagan and Thatcher inaugurated the neoliberal policies that promoted free trade and enabled offshoring. It will take several presidencies for this to happen. Such transformation will entail breaking existing economic relations, with hard consequences. By renouncing the role of the global consumer, the USA will likely enter in a position of conflict with other actors, including its allies, because the interests of the central subsystem will no longer coincide with those of the other subsystems.

11.5 Conclusions

This chapter has argued for a realist approach that views states as key actors in an interstate political system that coexists with multiple layers of social, economic, and cultural systems, each of them characterized by specific spatialities. The political system plays the key role of maintaining stability among the various systems by accommodating and managing the changes that occur over time in each system. When these changes are too profound or too rapid, the political system fails to maintain stability, leading to crises and conflicts of various nature and intensity, up to interstate wars. In such logic, wars are the ultimate expression of conflicts that can have various scalar impacts—from localized, peaceful protests, to civil wars, up to localized or system-wide interstate wars. Conflicts can remain isolated, or lead to chains of interrelated conflicts, such as those that started in Europe with the capitalist expansion of the nineteenth century that led to both world wars, that are arguably not fully resolved.

Following the logic of intersystemic change, the conflicts of the long twentieth century can be interpreted as the results of the tensions stemming from the various phases of globalization, driven by technological change leading to the spatial expansion of economic activities. After the shock of the Great Depression, the global system tried for a decade to regain its balance, which was eventually found through a new, gigantic conflagration, WW2. From a systemic perspective, the most important consequence of that war was the dissolution of the regionalized globalization that had emerged in the late nineteenth century, in which trade was, for the most part, constrained within the boundaries of colonial empires. In its place, the Cold War ushered duopolistic globalization, with the two integrated systems of the capitalist and communist worlds. The end of the Cold War ushered a new phase of unified globalization, which saw an unprecedented acceleration of trade and transfer of productive capacity from the systemic core of the USA and UK to the peripheries. Such process resembles closely historical developments in the Roman and Spanish empires, in which the emptying of the productive capacities was the precursor of imperial decadence. Similarly, the crises of 2008 and the current COVID crisis are exposing the vulnerabilities of the imperial center, and are accompanied by attempts to re-industrialize the core. However, such a process will be long, and fraught with conflicts of interests between the core—especially the USA that is trying to change its role of global consumer—and the rest of the system—that has a strong vested interest in keeping exporting to the core. In an earlier chapter of this book (Chap. 10) we have underlined the worrying similarities between the current situation and that which preceded the outbreak of the First and Second World Wars, even if the possible scenarios appear profoundly different (Huntington 2011). Whatever the case, the changing role of the USA in the global economy, as well as Brexit, indicates the beginning of a possible reverse of globalization, as countries will have to necessarily adapt to different trade flows. The mechanism of globalization may be coming at a standstill and we will have to wait a long time before it can start moving again. Like many other times in the past, the ocean of history has entered the ebb phase.

References

Aparicio Cabrera A (2013) Historia económica mundial 1870-1950. Economía Informa, 382, 99–115.

Battisti G (2011) Globalization reshaping regionalization. In: Proceedings of the 2nd conference on the adriatic forum, construction and deconstruction of nationalism and regionalism. The Long Journey to Europe. Academie Europeenne de Géopolitique, Montpellier, pp 17–22

Battisti G (2012) L'avvento del post-industriale nella ridefinizione degli equilibri geoeconomici [the emergence of the post-industrial in the redefinition of geoeconomic balances]. In: Dini F, Randelli F (eds) Memorie Geografiche, Oltre la globalizzazione: le proposte della Geografia Economica. Firenze University Press, Firenze, pp 25–34

Battisti G (2017) Iconographies of globalisation. Eur J Geogr 8(2):121–131

Fumian C (2003) Verso una società planetaria. Alle origini della globalizzazione contemporanea, Donzelli editore

Holzman FD (1982) Soviet military spending: assessing the numbers game. Int Secur 6(4):78–101

Huntington SP (2011) The clash of civilizations and the remaking of world order. Simon & Schuster

Lizza G (2009) Scenari geopolitici [geopolitical scenarios]. UTET, Torino

Luttwak E (1981) La grande strategia dell'impero romano [the grand strategy of the Roman Empire]. Rizzoli, Milano

Midlarsky MI (1990) Systemic wars and dyadic wars: no single theory. Int Interactions 16(3):171–181

Mulligan W (2017) The origins of the first world war. Cambridge University Press

Pinker S (2012) The better angels of our nature: why violence has declined. Penguin Group, USA

Pollio Salimbeni A (1999) Il grande mercato. Realtà e miti della globalizzazione [The great market. Myths and realities of globalization]. Bruno Mondadori, Milano

Todd E (2003) Dopo l'impero. La dissoluzione del sistema americano [After empire. The dissolution of TEH American system]. Tropea, Milano

Vallega, A. (1995) La regione, sistema territoriale sostenibile, Mursia

Wallerstein I (1978) Civilizations and modes of production: conflicts and convergences. Theory Soc 5(1):1–10

Wallerstein I (2015) Modern world-system in the longue duree. Routledge, London

Appendix: Gianfranco Battisti's Thematic Bibliography

The following list includes Gianfranco Battisti's publications concerning geopolitics and political geography. **Titles in bold** are the closest—in terms of subject matter and methodology—to this book's chapters.

Books

Battisti, G. (1979) Una regione per Trieste. Studio di geografia politica ed economica [A region for Trieste. An essay on political and economic geography], *Trieste,* Istituto di Geografia, Facoltà di Economia e Commercio, Università di Trieste, pp. 279.

(1984) La collaborazione del Friuli-Venezia Giulia con la Jugoslavia e le prospettive di sviluppo [The economic cooperation between Friuli Venezia Giulia and Jugoslavia, and the perspectives of development], Excerpt from: Atti 8° Convegno Scientifico Internazionale Alpe-Adria (Graz, 7–8 ottobre 1982): Interventi selezionati per la promozione dei rapporti economici nelle Regioni centrali della Comunità di lavoro Alpe-Adria. Incontro tra teoria e pratica, Facoltà di Economia e Commercio, Università di Trieste, pp. 107–223.

(1988) L'economia di frontiera in una regione problema [The border economy in a problem-region], Trieste, I.R.S.E.T., Pubbl. n. 5, pp. 70.

(1993) Studies on geopolitical reconstruction, Trieste, I.R.S.E.T., 1993, pp. 23, tavv.

(1996) Atti del Convegno di Studi in onore di Giorgio Valussi, Trieste, 6–7 febbraio 1992 [Proceedings of conference in honour of Giorgio Valussi, Trieste, February 6–7, 1992], (a cura di G. Battisti e P. Nodari), Trieste, Università degli Studi di Trieste - Dipartimento di Scienze Geografiche e Storiche, vol. II, pp. 206.

(2002) Un pianeta diviso. Contributi alla geografia dei popoli e dei confini [A divided planet. Essays on the geography of peoples and borders], (a cura di), Trieste, Università degli Studi di Trieste – Dipartimento di Scienze Geografiche e Storiche, 2002, (2, Sr. Geografia), pp. 266.

(2002) Les Habsbourg 1848–1916. Moments d'une dynastie [The Habsburgs 1848–1916. Moments of a dynasty], Firenze, Alinari Idea, 2002 (con Z. Ciuffoletti e G. Favroud).

(2008) Dalla dissoluzione dei confini alle Euroregioni. Le sfide dell'innovazione didattica permanente. [From the dissolution of borders to Euroregions. The challegences of permanent didactic innovation], Atti 51° Convegno Nazionale AIIG (Trieste, 15–22 agosto 2008), Firenze, Le Lettere, vol. I, pp. 268 (a cura di), 2011.

Articles in Scientific Journals

(1978) La "questione di Trieste" e il suo inquadramento regionale [The Trieste question in a regional perspective], in Boll. Soc. Geogr. Ital., Sr. 10, 7, pp. 23–40.

(1981) I movimenti "turistici" al confine italo-jugoslavo. Metodologie per la ricerca [Tourist trips across the Italian/Yugolav border. Research methodologies], in Quaderni del C. S. E. Vanoni di Trieste, 15, pp. 1–13.

(1989) L'economia di frontiera [Border economy], in Friuli Venezia Giulia: Regione Problema. Aggiornamenti scientifici e didattici, Atti del XXXII Conv. Naz. dell'A.I.I.G. (a cura di G. Valussi), Geografia nelle Scuole, 1989, pp. 41–49.

C. Sellar and G. Battisti, *Geopolitical Perspectives from the Italian Border*, Historical Geography and Geosciences, https://doi.org/10.1007/978-3-031-26044-5

(1994) Pulizia etnica, ovvero la negazione della geografia [Ethnic cleansing, or the denial of geography], Geografia ed educazione allo sviluppo, (dossier a cura di G. De Vecchis), in Il Mondo domani, 2, pp. 24–25.

(1997) Un mondo di Jugoslavie [A world of Yugoslavias], in Quaderni del Centro Studi Economico-Politici "Ezio Vanoni", 1–3, pp. 9–11.

(2003) Verso una struttura imperiale? Il riaffacciarsi di vecchi concetti nella prospettiva della nuova Europa [Towards an imperial structure? The reemergence of old concepts in the perspective of the new Europe], in Boll. Soc. geogr. Ital., 2, pp. 355–374.

(2010) Demografia, politica ed economia. Un'intreccio che modella la geografia del pianeta [Demograpy, politics, and economy. An intertwining that shapes the geography of the planet], in Liguria Geografia, 12, 9, pp.3–5.

(2011) Globalizzazione e schiavitù [Globalization and slavery], in Geografia, 34, pp. 3–9.

(2011) L'Europa come idea e progetto [Europe as an idea and a project], Bollettino dell'Osservatorio Internazionale Cardinale Van Thuan sulla Dottrina Sociale della Chiesa, 8, 1, pp. 9–15.

(2013) Capitali senza confini. L'antimondo dei paradisi fiscali e i flussi finanziari internazionali [Capital with no borders. The antiworld of financial heavens and international financial flows], in: Bollettino dell'Osservatorio internazionale Card. Van Thuan sulla Dottrina sociale della Chiesa, 9, pp. 103–107.

(2014) Offshoring and Financial Markets, in Economy of Region, 10, 2, pp. 150–160.

(2014) Towards a Geography of Financial Relationships, European Journal of Geography, 5, 2, pp. 18–26.

(2014) Governing globalisation. The energy debate between nature and macroeconomic issues, Semestrale di studi e ricerche di Geografia, 27, 1, pp. 5–21.

(2015) La geopolitica delle religioni nell'epoca del ritorno del paganesimo [The geopolitics of religions in the age of the return of paganism], Bollettino dell'OsservatorioInternazionale Cardinale Van Thuan sulla Dottrina Sociale della Chiesa, 11, 4, pp. 145–151.

(2017) Iconographies of Globalisation, European Journal of Geography, 8, 2, pp. 121–131.

(2017) Le aree interne: un'interpretazione in chiave geopolitica [Inner areas: a geopolitical interpretation], Geotema, 21, pp. 85–88.

(2019) Islam: un problema geopolitico? [Islam: a geopolitical issue?], Bollettino dell'Osservatorio Internazionale Cardinale Van Thuan sulla Dottrina Sociale della Chiesa, 15, 2, pp. 60–69.

(2019) **Fine della globalizzazione o tramonto dell'Occidente? [End of globalisation or decline of the West?], in Boll. Soc. Geogr. It., Sr. 14, 2, 1, pp. 129–138**.

(2020) Le transizioni energetiche: un inquadramento geopolitico [Energy transitions: a geopolitical framework], in QuaderniCIRD. Rivista del Centro Interdipartimentale per la Ricerca Didattica dell'Università di Trieste/ Journal of the Interdepartmental Center for Educational Research of the University of Trieste, 21, pp. 1–15.

(2022) La questione energetica nel terzo millennio [The Energy Issue in the Third Millenium], forthcoming in Geotema.

Book Chapters

(1978) Ondate migratorie anomale nell'esperienza italiana: i profughi [Anomalous migratory waves in the Italian experience: the refugees], in Italiani in movimento (a cura di G. Valussi), Commissione di studio sui fenomeni migratori dell'A.Ge.I., Pordenone, pp. 217–222.

(1979) Tourism and Borders between Italy and Yugoslavia: a Methodology for the Research, in Seminar für Wirtschats- und Sozialgeographie, J. W. Goethe Universität, Frankfurt/Main, Heft 31, pp.215–229.

(1983) Cosmopolitismo e municipalismo quali fattori di identificazione della comunità triestina [Cosmopolitism and municipalism as identity factors of the Trieste community], in Atti 23° Congr. Geogr. Ital. (Catania, 9–13 maggio 1983), A.Ge.I./Istituto di Geografia, Facoltà di Lettere e filosofia, Università di Catania, pp. 269–278.

(1986) Le relazioni frontaliere sul confine tra Friuli-Venezia Giulia e Slovenia 1982–1985. Tra crisi ed evoluzione [Crossborder relations between *Friuli-Venezia Giulia* and Slovenia 1982–1985], in Atti 10° Convegno Scientifico Internazionale Alpe-Adria (Trieste, 24–25 ottobre 1986), Trieste, Università degli Studi di Trieste, Facoltà di Economia e Commercio, pp. 61–103.

(1988) Trieste: la difficile trasformazione di un crocevia di traffici [Trieste: the difficult metamorphosis of a traffic crossroads], in L'economia di frontiera in una regione problema, Trieste, I.R.S.E.T., 5, pp. 60–66.

(1989) Quando la strategia modella la carta geopolitica: le Province illiriche [The Illyrian provinces: when the strategy shapes the geopolitical maps], in: Trieste, Nodier e le Province Illiriche- Atti del Convegno - Trieste, 17 novembre 1987, (a cura di G. Casa), Trieste, SIDEF - Provincia di Trieste, pp. 39–46.

(1993) Sistema mondo, [world system], Centri e periferie [Centers and peripheries], Abitare il pianeta [Living the planet], in Il mondo. Geografia economica. Corso per le scuole superiori, Milano, Garzanti, pp. 1–23; 31–35; 53–56.

(1995) La nuova geografia dell'Est europeo. Spunti per una riflessione [The new geography of Eastern Europe. Ideas for reflection], in Percorsi geografici 1992–95 (a cura di G. Galliano), Genova, A.I.IG.—Sez. Liguria, pp. 73–88.

(1996) Realtà e problemi delle aree di confine. Per un'analisi geografica delle aree di frontiera [Problems and realities of border areas. For a geographical analysys of border areas], in Atti del Convegno di studi in onore di Giorgio Valussi (Trieste, 6–7 febbraio 1992), (a cura di G. Battisti e P. Nodari), Università degli Studi di Trieste - Dipartimento di Scienze Geografiche e Storiche, Trieste, volume 2, pp. 7–106.

(1996) The Frontier Problems of the new Europe. Guide-lines for the Round Table and Contributions to the Section, A.Ge.I., 27° Italian Geographical Congress, Directions in Italian Geography. The Geography of Challenge and Change, Bologna, Pàtron, pp. 22–24.

(1999) L'immigrazione straniera in Italia» [Foreign immigration to Italy], in Italia crocevia di genti. Immigrazione al positivo: la nascita di una cultura multietnica. Atti del Convegno: Cassino, 29–30 aprile 1997, (a cura di G. Arena, A. Riggio, P. Visocchi), Perugia, Università di Cassino, pp. 49–59.

(2000) La géopolitique de l'Adriatique, un résumé historique [Adriatic geopolitics, an historical excursion], in A.-L. Sanguin, Mare Nostrum, dynamiques et mutations géopolitiques de la Méditerranée, Paris, L'Harmattan, pp. 265–274.

(2000) Di crisi in crisi. Il travaglio di un'economia [From crisis to crisis. The travail of an economy], in Il secolo Asburgico 1848–1914, Fotografie di un impero, (a cura di G. Ciuffoletti), Firenze, Fratelli Alinari, pp. 22–36.

(2000) La Dalmazia. Natura e funzioni di un organismo geografico», in Contributi alla conoscenza della Storia e della cultura dell'Istria, di Fiume e della Dalmazia (a cura di S. Cattalini), Udine, ANVGD, pp. 31–47.

(2001) Applicazione di un modello ARIMA per lo studio dei movimenti transfrontalieri nella provincia di Trieste [Transborder flows in Trieste district described by ARIMA methods], in La Geografia delle sfide e dei cambiamenti – Atti XXVII CGI (Trieste, 1996), Bologna, Patron, 2001, pp. 539–551 (con A. Favretto).

(2001) The Gulf Wars. A geopolitical interpretation, in On the Centenary of Ratzel's Politische Geographie. Europe Between Political Geography and Geopolitics, eds. M. Antonsich, V. Kolossov, M. Paola Pagnini, Roma, Soc. Geogr. Ital., 2001, pp. 447–459.

(2002) The logic of empire. Old and new concepts in the European geopolitical reconstruction, in L. Buzzetti, E. Bullado (eds.), Geographical renaissance at the dawn of the millennium. The Italian perspective, Roma, Società Geografica Italiana, pp. 99–114.

(2001) Applicazione di un modello ARIMA per lo studio dei movimenti transfrontalieri nella provincia di Trieste [Transborder flows in Trieste district described by ARIMA methods], in La Geografia delle sfide e dei cambiamenti – Atti

27° CGI (Trieste, 1996), Bologna, Patron, pp. 539–551 (in coll. con A. Favretto).

(2002) Tra confine e frontiera: la regione "mobile" [Among border and frontier: the mobile region], in Un pianeta diviso. Contributi alla geografia dei popoli e dei confini, (a cura di), Trieste, Univ degli Studi – Dipartimento di Scienze Geografiche e Storiche, (2, Sr. Geografia), pp. 101–114.

(2006) La società multiculturale, ovvero la fine di un illusione [The multicultural society as the end of an illusion], in Scritti per Alberto Di Blasi (a cura di G. Campione, F. Farinelli, C. Santoro Lezzi), Bologna, Pàtron, pp. 173–180.

(2006) L'Europa che non sarà [The Europe that will not be], in Scritti in onore di Roberto Bernardi (a cura di S. Salgaro), Bologna, Pàtron, vol. I, pp. 439–446.

(2007) Lo scontro delle civiltà – una rivisitazione [The clash of civilizations. A reinterpretation], in Geografia. Dialogo tra generazioni. Atti 29° Congresso Geografico Italiano: Palermo, settembre 2000 (a cura di A. Di Blasi), Bologna, Pàtron, vol. I, pp. 303–306.

(2009) La rinascita della Russia [The rebirth of Russia], in Una vita per la Geografia Scritti in ricordo di Piero Dagradi (a cura di C. Cencini, L. Federzoni, B. Menegatti), Bologna, Pàtron, pp.315–324.

(2011) La lunga strada verso l'Euroregione [The long way towards the Euroregion], in Dalla dissoluzione dei confini alle Euroregioni. Le sfide dell'innovazione didattica permanente. Atti 51° Convegno Nazionale AIIG: Trieste, 15–22 agosto 2008 (a cura di), Firenze, Le Lettere, vol. I, pp. 13–32.

(2011) Which Kind of Europe. The Identity of EU as outlined in the Constitutional Project, in Europe seen from here and elsewhere. Proceedings of EuroBroadMap Conference: Rouen, 1 & 2 December 2011 (C. Didelon and C. Grasland, eds.), Rouen, CCSd, pp. 28–31.

(2011) Globalization reshaping regionalization, in Proceedings of the 2nd Conference of the Adriatic Forum. Construction and Deconstruction of Nationalism and Regionalism: A Long Journey to Europe (M. P. Pagnini, ed.), Montpellier, Academie Europeenne de Géopolitique, pp. 17–22.

(2011) Gorizia Nova, un'eurocittà al confine tra Italia e Slovenia [Gorizia Nova a Euro-city on the border between Italy and Slovenia], in A Pasquale Coppola. Raccolta di scritti (a cura di L: Viganoni), Roma, Società Geografica. Italiana, Memorie 79, vol. I, pp. 43–56.

(2012) L'avvento del post-industriale nella ridefinizione degli equilibri geoeconomici emergence of the post-industrial in the redefinition of geo-economic balances], in Oltre la globalizzazione. Le proposte della geografia economica (a cura di F. Dini e F. Randelli), Firenze, Società di Studi Geografici - Firenze University Press, pp. 25–34. Memorie Geografiche - N.S., n. 9.

(2013) Movimenti migratori o nuova tratta? [Migratory movements or a new slave trade?], in Migrazioni di ieri e di oggi. Verso una nuova società tra integrazione, sviluppo e globalizzazione (a cura di F. Krasna), Bologna, Pàtron, pp. 202–211.

(2013) Mercato globale e valute regionali. Un equilibrio conflittuale [Global markets and regional currencies. A conflictual balance], in Oltre la globalizzazione. Prossimità/Proximity (a cura di C. Capineri, F. Celata, D. de Vincenzo, F. Dini, F. Randelli, P. Romei), Firenze, Società di Studi Geografici - Firenze University Press, pp. 81–90. Memorie Geografiche - N.S., n. 11.

(2014) Equilibri in mutamento nell'Europa del terzo millennio. Il caso del confine orientale d'Italia [Changing balances in Europe in the 3rd millenium. The case of Italian Eastern border], in Atti 31° Congresso Geografico Italiano: Milano, 11–13 giugno 2012 (a cura di G. Scaramellini e E. Mastropietro), Milano, Mimesis, Vol. I, pp. 277–287.

(2014) Gli inizi dell'industria petrolifera iraniana: risorse naturali e determinismo ambientale" [The beginnings of Iranian oil industry: natural resources and environmental determinism], in Un lungo viaggio nella geografia della Sardegna. Studi in onore di Antonio Loi (a cura di M. Tanca), Bologna, Pàtron, pp. 101–109.

(2014) Geopolitical Dislocation as a product of the Financialisation of Economy. Lessons from Europe, in A Geographical Approach to the European Financial Crisis. Challenges and Policy Agenda (M. G. Lucia, L. M. Rizzo eds.), Roma, Aracne, pp. 73–84.

(2014) La legge rango-dimensione quale metodologia preliminare all'analisi della resilienza. Il caso del Litorale Austro-Illirico [The rank-size rule as a preliminary tool for the study of resilience. The case of Austrian-Illyrian Littoral], in Oltre la globalizzazione. Resilienza/Resilience (a cura di C. Capineri, F. Celata, D. de Vincenzo, F. Dini, F. Randelli, P. Romei), Firenze, Società di Studi Geografici —Firenze University Press, pp. 193–198 Memorie Geografiche - N.S. n. 12.

(2015) The identity of the European Union as outlined in the Constitutional Project, in Verso un nuovo paradigma geopolitico. Raccolta di scritti in onore di Gianfranco Lizza (a cura di M. Marconi e P. Sellari), Ariccia (Roma), Aracne, Vol. 2, pp. 457–468.

(2015) Rotture geopolitiche e cambiamenti paradigmatici: il ritorno dell'economia politica [Geopolitical ruptures and paradigm shifts: the return of political economy], in Oltre la globalizzazione. Conflitti/Conflicts (a cura di C. Capineri, D. De Vincenzo, F. Dini, M. Lazzeroni, F. Randelli), Firenze, Società di Studi Geografici—Firenze University Press, pp. 165–171. Memorie Geografiche—N.S. n. 13.

(2016) A cent'anni dalla "Grande guerra": scenari geopolitici a confronto [Comparing geopolitical scenarios in the centenary of the Great War], in Studi in onore di E. Paratore. Spunti di ricerca per un mondo che cambia (a cura di L. Romagnoli), Roma, EDIGEO, pp. 937–948.

(2017) La moltiplicazione degli operatori creditizi e l'emergere del sistema bancario ombra [The fragmentazione of credit sector and the emergence of the shadow banking system], in (S) radicamenti (a cura di E. Dansero, M. G. Lucia, U. Rossi e A. Toldo), Firenze, Società di Studi geografici—Firenze University Press, pp. 589–595. Memorie Geografiche—N. S., n. 15.

(2017) Europa. Le molte ragioni di una crisi epocale [Europe. The many reasons for an epochal crisis], in Europa: la fine delle illusioni, Nono Rapporto sulla Dottrina sociale della Chiesa nel mondo (a cura di G. Crepaldi e S. Fontana), Siena, Osservatorio Internazionale Card. Van Thuan sulla Dottrina sociale della Chiesa - Cantagalli, pp. 123–145.

(2018) Dopo la globalizzazione. Verso un ritorno alle gabbie d'oro? [After globalisation. Towards a return to golden fetters?], Barriere/ Barriers (a cura di M. Fuschi), Firenze, Società di Studi Geografici—Firenze University Press, 2018, pp. 57–66. Memorie Geografiche—N. S. n. 16.

(2019) La ciclicità degli assetti geopolitici come portato delle dinamiche delle strutture spaziali" [The cyclical nature of geopolitical assets as a result of spatial dynamics], in L'apporto della Geografia tra rivoluzioni e riforme. Atti del XXXII Congresso Geografico Italiano: Roma, 7–10 giugno 2017 (a cura di F. Salvatori), Roma, A.Ge.I., pp. 1091–1098.

(2019) Shadow banking: a geographical interpretation, in S. Grandi, C. Sellar, J. Jafri (eds.), Geofinance between political and financial geographies, London, Edward Elgar, pp. 49–64.

(2020) I popoli come obiettivo e come strumento. Una prospettiva geopolitica" [Peoples as goals and tools. A geopolitical perspective], in Popoli, Nazioni, Patrie: tra natura e artificio politico. Undicesimo Rapporto sulla Dottrina sociale della Chiesa nel mondo (a cura di G. Crepaldi e S. Fontana), Siena, Osservatorio Internazionale Card. Van Thuan sulla Dottrina sociale della Chiesa - Cantagalli, pp. 43–64.

(2020) Trieste: many dimensions of a disputed port-city- At the Crossroads of Spaces and Times, in Gerry O'Reilly (ed.), Places of Memory and Legacies in an Age of Insecurities and Globalization, Berlin, Springer, 2020, pp. 265–277.

(2020) Esaurimento delle risorse? Il caso del petrolio", in Dodicesimo Rapporto sulla Dottrina sociale della Chiesa nel mondo Ambientalismo e globalismo: nuove ideologie politiche (a cura di R. Cascioli, G. Crepaldi, S. Fontana), Siena, Osservatorio Internazionale Card. Van Thuan sulla Dottrina sociale della Chiesa - Cantagalli, pp. 113–131.

(2021) Il modello cinese e la nuova partita a scacchi della geopolitica mondiale [The Chinese model and the new chess game of world geopolitics], in Il modello cinese: capital-socialismo del controllo sociale Tredicesimo Rapporto sulla Dottrina sociale della Chiesa nel mondo (a cura di R. Cascioli, G. Crepaldi, S. Fontana), Siena, Osservatorio Internazionale Card. Van Thuan sulla Dottrina sociale della Chiesa - Cantagalli, 2021, pp. 61–83. French translation: (2022) Le modèle chinois et la nouvelle partie d'échecs de la géopolitique mondiale in Les noveaux habits deu totalitarisme, Paris, Editions de L'homme noveau, pp. 67–88.

(2022—expected) Sapere è potere. Il ruolo dell'informazione sul mercato degli idrocarburi [Knowledge is power. The role of information on the hydrocarbon market], in Memorie Geografiche—N.S. n., Firenze, Società di Studi Geografici—University Press, 2022, (in press).

(2023—expected) Le mafie istituzionali come oggetto di indagine geoeconomica" ["Institutional mafias'as a subject of geoeconomic investigation], Oltre la Globalizzazione 2021, Catene/Chains, Memorie Geografiche—N. S. n., Firenze, Società di Studi Geografici—Firenze University Press, 2023 (in press).

(2023—expected) Scenari ambientali e strategie energetiche. Realtà, rappresentazione, trasformazione [Environmental scenarios and energy strategies. Reality, representation, transformation], in Geografie in movimento. Atti 33° Congresso Geografico Italiano (Padova, 8–13 settembre 2021), in Moving geographies. Proceedings of the 33rd Italian Geographical Congress 2021 (in press).

Index

A
Administrative units, 9, 38, 47–51, 55, 56, 59, 60, 64, 65
Anglo-American (geography), 5, 13, 14, 18–20
Austria, 8, 33, 34, 47, 49, 51, 52, 73, 82, 83, 86, 91, 92, 96

B
Bonetti (Eliseo), 27, 28, 35, 37, 39, 40
Border, 4–6, 8–10, 15, 19, 34–36, 38–41, 48–52, 55, 57, 59–64, 66, 67, 69–75, 81, 83, 85, 92, 93, 97, 99, 109, 113, 121–124, 126
Borderlands, 5, 35, 38, 39, 70, 71, 73, 74
Bretton Woods, 106, 107
Brexit, 95, 113, 124, 128, 129

C
China, 95, 100, 104, 105, 108–117, 127, 128
Christaller (Walter), 6, 9, 27, 50, 51, 74, 75
Cold War, 3–6, 8, 9, 20, 33, 40, 52, 60, 62, 79, 80, 90, 93, 105, 116, 126, 128
Conflicts, 8–10, 33, 35, 38, 51, 56, 59, 79, 80, 92, 93, 95–97, 100, 101, 104, 106, 107, 109–112, 114–118, 118, 121–124, 128, 129
Core, 5, 6, 17, 28, 30, 63, 64, 85, 87, 96, 99, 100, 110, 128, 129
Croatia, 38, 42, 47–49, 51, 52, 57, 60, 74, 75, 80–84, 96

D
Disequilibrium, 10, 121

E
Economic development, 9, 52, 69, 125, 127
Economic geography, 5, 7, 8, 13, 14, 16, 18, 20, 21, 27–29, 33, 36, 55
Empire, 8, 28, 34, 47–49, 52, 56, 57, 60, 69, 71, 85, 90–93, 95, 96, 101, 103, 106, 109–111, 123, 126–128
Equilibrium, 8, 121, 122
Euro, 10, 59, 60, 94, 96, 103, 105, 116
Euro (currency), 10, 59, 94, 96, 97, 101, 103, 105–107, 116
European Commission, 94
European integration, 5, 42, 94, 97
Europeanization, 6, 7
European Union (EU), 4, 9, 10, 20, 33, 42, 49, 51, 59, 60, 62–66, 70, 79, 80, 89, 93–101, 103, 105, 106, 108, 109, 111–116, 118, 124, 128

G
Geoeconomics, 6, 36
Geographical tradition, 4, 6, 8, 13, 21, 42, 80
Geographic thought, 13, 14, 20
Geographies of geography, 4
Geopolitics, 3–11, 33, 36, 37, 39–42, 59, 66, 67, 69, 79, 80, 82, 87, 89, 101, 103, 121
Germany, 27, 28, 51, 80, 82, 85, 87, 92–94, 96–101, 104, 109–111, 124, 126, 127
Globalization, 3, 5, 7, 9, 10, 93, 100, 101, 116, 121, 124–129
Gorizia, 9, 35, 38, 40, 53, 59–67, 75
Great Depression, 104, 107, 126, 128

H
Hegemony, 4, 6, 8, 10, 74, 103, 104, 110, 116, 124, 127
Historical geographies, 7, 8, 10, 19, 20, 30, 37, 42, 59, 69, 79, 89, 121
Human geography, 4, 6, 7, 13, 19, 29, 36, 39, 40, 47, 69, 79

I
Industry, 3, 6, 7, 13, 14, 17–21, 26, 74, 99, 103, 107, 109, 118, 127
Inland areas, 70
Inner areas, 42
Innovation, 7, 13–21, 27, 28, 118
Institute of Geography, 25, 26, 28–30, 34, 35
Institutions, 6, 14–21, 55, 56, 61, 71, 82, 94, 101, 107, 122
International (geography), 3–7, 10, 13, 28, 40, 50, 79
Iron Curtain, 5, 8, 33, 40
Istrian Peninsula, 35, 50, 52

C. Sellar and G. Battisti, *Geopolitical Perspectives from the Italian Border*, Historical Geography and Geosciences, https://doi.org/10.1007/978-3-031-26044-5

Italy, 5, 7–9, 13, 14, 16–20, 25–30, 33, 35–38, 47–52, 55, 56, 59–62, 64, 69, 71, 73–75, 80–85, 86, 87, 90, 92, 93, 97–99, 104, 122, 124

K

Knowledge production, 3, 4, 6, 14, 15, 18, 19

L

Landscape studies, 7, 8, 20, 28, 33, 35, 38, 50
Littoral, 47–53, 55–57, 59, 65, 74
Loesch, 6, 9

M

Marginality, 69, 70, 71, 73, 74

N

Nation-states, 10, 57, 70, 90–94, 100, 113

O

Organizations, 8, 13–15, 18, 34, 38, 39, 42, 47, 52, 55–57, 62, 91, 93–95, 98, 114, 116, 126, 127

P

Path dependencies, 6, 7, 14
Peace, 9, 35, 38, 48, 50, 59, 62, 66, 98, 101, 104, 106, 114, 116, 118, 121
Periphery, 70, 72, 80, 87, 128
Political geography, 14, 27, 33, 36, 37, 39–41
Postmodern, 13, 127

Q

Quantitative revolution, 6–8, 13, 18–20, 28, 29, 39, 47

R

Rank Size Rule (RSR), 47, 50–53, 55, 56
Realism, 6, 10, 20, 121
Realist approaches, 10, 20, 80, 103, 128
Resiliency, 48, 71
Ritter (Carl), 30
Roletto (Giorgio), 35–40, 42
Russia, 91, 92, 95–97, 100, 101, 109–118, 123, 124, 127

S

Scale, 9, 10, 15–18, 30, 38, 42, 55, 62, 67, 81, 82, 90, 93, 103, 110, 111, 121, 123, 124, 126, 127
Seigniorage, 10, 103, 105, 106
Slovenia, 9, 38, 42, 47–49, 52, 57, 59–66, 74, 75, 80–84, 96
Sovereignty, 5, 6, 94
Soviet Union, 40, 80, 85, 98, 104, 105
Spatial relations, 10, 121
Stock exchange, 107
Storper (Michael), 7, 14, 16–20
Structuration theory, 15, 16
Subsystem, 10, 122–124, 128
System, 6–8, 10, 11, 13, 16, 17, 19, 41, 48, 50–52, 55–57, 62, 64, 65, 80, 85, 87, 90, 92, 100, 101, 104, 106–108, 112, 114–118, 121–129

T

Territorialization, 5, 47, 48, 59, 66
Territorial states, 7, 10, 89, 121
Territorial structures, 40, 47, 48, 93
Territorial trap, 7, 10, 89
Treaty of Osimo, 9, 38, 39, 81
Trieste, 3–5, 7–9, 14, 16, 18–21, 25–28, 33, 35–42, 48, 50–53, 55, 56, 59, 60, 62, 64, 66, 69, 74, 75, 80, 81, 83
20th Century, 6–8, 33, 47, 48, 56, 91, 92, 103, 114, 124, 126, 128

U

Underdevelopment, 9, 69, 70, 73
United States, 10, 99, 103, 105, 108, 112, 113, 117, 126
Untraded interdependencies, 6, 13, 14, 16, 17, 20
Urban areas, 9, 50, 59–61, 63, 64, 75, 85
Urban hierarchy, 9, 47
USA, 96, 97, 103–105, 107, 108, 112–114, 124, 127, 128
US dollar, 10, 103, 106–108

V

Valussi (Giorgio), 35, 36, 38–41, 50, 52, 73, 75, 83
Venezia Giulia, 38, 48–50, 55, 56, 62, 65, 74, 83

W

War, 8, 9, 17, 33, 35, 36, 38, 57, 59, 61, 65, 71, 74, 79, 80, 82, 84, 85, 92, 95–97, 99–101, 103–111, 113, 114, 116–118, 121–129
Worlds of production, 13, 16, 17, 20
World War 1 (WW1), 5, 8–10, 33, 35, 48, 49, 51, 52, 55, 56, 91, 92, 103, 110, 111, 126
World War 2 (WW2), 5, 8, 9, 33–38, 48–50, 52, 55, 56, 59–61, 80, 92, 94, 100, 103, 104, 107, 122, 126, 128

Y

Yugoslavia, 8, 9, 33, 35, 38, 40, 48, 50, 52, 55, 57, 59, 60, 62, 65, 75, 79–82, 84, 122